# METHODS IN MOLECULAR BIOLOGY™

*Series Editor*
**John M. Walker**
**School of Life Sciences**
**University of Hertfordshire**
**Hatfield, Hertfordshire, AL10 9AB, UK**

For further volumes:
http://www.springer.com/series/7651

# Liver Stem Cells

## Methods and Protocols

Edited by

## Takahiro Ochiya

*National Cancer Center Research Institute, Tokyo, Japan*

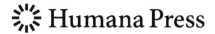 Humana Press

*Editor*
Takahiro Ochiya, Ph.D.
Chief, Division of Molecular and Cellular Medicine
National Cancer Center Research Institute
5-1-1, Tsukiji, Chuo-ku, Tokyo 104-0045
Japan
tochiya@ncc.go.jp

ISSN 1064-3745          e-ISSN 1940-6029
ISBN 978-1-61779-467-4      e-ISBN 978-1-61779-468-1
DOI 10.1007/978-1-61779-468-1
Springer New York Dordrecht Heidelberg London

Library of Congress Control Number: 2011942927

Printed on acid-free paper

Humana Press is part of Springer Science+Business Media (www.springer.com)

# Preface

## A Brief Outline of the Aims and Target Audience of *Liver Stem Cells*

The role of a putative stem cells and liver-specific stem cell in regeneration and carcinogenesis is reviewed in this book.

There is increasing evidence that there is a liver stem cell that has the capacity to differentiate into parenchymal hepatocytes or into bile ductular cells. These stem cells may be activated to proliferate after severe liver injury or exposure to hepatocarcinogens. Stem cell replacement strategies are therefore being investigated as an attractive alternative approach to liver repair and regeneration. In this book, we focus on recent preclinical and clinical investigations that explore the therapeutic potential of stem cells in repair of liver injuries. Several types of stem cells, such as embryonic stem (ES) cells, induced pluripotent stem (iPS) cells, haematopoietic stem cells, and mesenchymal stem cells, can be induced to differentiate into hepatocyte-like cells in vitro and in vivo. Stem cell transplantation has been shown to significantly improve liver function and increase survival in experimentally induced liver-injury models in animals. Furthermore, several pilot clinical studies have reported encouraging therapeutic potential of stem cell-based therapies. This book consists of five main categories: (1) Several hepatic progenitor cells; (2) Hepatic differentiation from stem cells; (3) Bile ductal cell formation from stem cells; (4) Liver stem cells and hepatocarcinogenesis; and (5) Application of liver stem cells for cell therapy. All these current topics shed light on stem cell technology which may lead to the development of effective clinical modalities for human liver diseases.

I believe this book will become the gold standard on this topic and will be widely distributed and read by people in many scientific fields, such as cellular biology, molecular biology, tissue engineering, liver biology, cancer biology, and stem cell therapy.

*Tokyo, Japan*                                                                                           *Takahiro Ochiya*

# Contents

PART III   BD FORMATION FROM STEM CELLS

PART IV   LIVER STEM CELLS AND HEPATOCARCINOGENESIS

PART V   APPLICATION OF LIVER STEM CELLS FOR CELL THERAPY

# Contributors

AGNIESZKA BANAS • *Laboratory of Molecular Biology, Institute of Obstetrics and Medical Rescue, University of Rzeszów, Faculty of Medicine, Rzeszow, Poland*

ALAIN CHAPEL • *IRSN, DRPH/SRBE/LTCRA, CEDEX 92262, France*

YOCK YOUNG DAN • *Department of Medicine, Yong Loo Lin School of Medicine, National University of Singapore, Singapore*

MARI DEZAWA • *Department of Stem Cell Biology and Histology, Tohoku University Graduate School of Medicine, Sendai, Japan*

F. EVENOU • *Laboratoire Matière et Systèmes Complexes (MSC), Bâtiment Condorcet, Université Paris Diderot, Paris 7, France*

SABINE FRANÇOIS • *IRSN, DRPH/SRBE/LTCRA, CEDEX 92262, France*

T. FUJII • *Institute of Industrial Science, University of Tokyo, Tokyo, Japan*

LUC GAILHOUSTE • *Division of Molecular and Cellular Medicine, National Cancer Center Research Institute, Tokyo, Japan*

LAURA GOMEZ-SANTOS • *Metabolomics Unit, CIC bioGUNE, Technology Park of Bizkaia, Bizkaia, Basque Country, Spain*

KEITARO HAGIWARA • *Division of Molecular and Cellular Medicine, National cancer Center Research Institute, Tokyo, Japan*

M. HAMON • *Department of Mechanical Engineering, Auburn University, Auburn, AL, USA*

SURADEJ HONGENG • *Department of Pediatrics, Faculty of Medicine, Ramathibodi Hospital, Mahidol University, Bangkok, Thailand*

H. HUANG • *Okami Chemical Industry Co. Ltd, Kyoto, Japan*

NORIHISA ICHINOHE • *Department of Tissue Development and Regeneration, Research Institute for Frontier Medicine, Sapporo Medical University School of Medicine, Sapporo, Japan*

MITSURU INAMURA • *Department of Biochemistry and Molecular Biology, Graduate School of Pharmaceutical Science, Osaka University, Osaka, Japan*

TETSUYA ISHIKAWA • *Core Facilities for Research and Innovative Medicine, National cancer Center Research Institute, Tokyo, Japan*

JUNFANG JI • *Laboratory of Human Carcinogenesis, Bethesda, MD, USA*

SHUICHI KANEKO • *Center for Liver Diseases, Kanazawa University Hospital, Kanazawa, Japan; Department of Gastroenterology, Kanazawa University Graduate School of Medical Science, Kanazawa, Japan*

KENJI KAWABATA • *Laboratory of Stem Cell Regulation, National Institute of Biomedical Innovation, Osaka, Japan*

MASAAKI KITADA • *Department of Stem Cell Biology and Histology, Tohoku University Graduate School of Medicine, Sendai, Japan*

JUNKO KON • *Department of Tissue Development and Regeneration, Research Institute for Frontier Medicine, Sapporo Medical University School of Medicine, Sapporo, Japan*

N. KOJIMA • *Institute of Industrial Science, University of Tokyo, Tokyo, Japan*

YASUMASA KURODA • *Department of Stem Cell Biology and Histology, Tohoku University Graduate School of Medicine, Sendai, Japan*

JOSE MARIA MATO • *CIC bioGUNE, Technology Park of Bizkaia, Bizkaia, Basque Country, Spain*

MARIA LUZ MARTINEZ-CHANTAR • *CICbioGUNE, Metabolomics Unit, Bizkaia, Basque Country, Spain*

TOSHIHIRO MITAKA • *Department of Tissue Development and Regeneration, Research Institute for Frontier Medicine, Sapporo Medical University School of Medicine, Sapporo, Japan*

ATSUSHI MIYAJIMA • *Laboratory of Cell Growth and Differentiation, Institute of Molecular and Cellular Biosciences, The University of Tokyo, Tokyo, Japan*

K.P. MONTAGNE • *Institute of Industrial Science, University of Tokyo, Tokyo, Japan*

HIROYUKI MIZUGUCHI • *Department of Biochemistry and Molecular Biology, Graduate School of Pharmaceutical Sciences, Osaka University, Osaka, Japan*

MOUBARAK MOUISEDDINE  *IRSN, DRPH/SRBE/LTCRA, CEDEX 92262, France*

YUJI NISHIKAWA • *Division of Tumor Pathology, Department of Pathology, Asahikawa Medical University, Asahikawa, Japan*

T. NIINO • *Institute of Industrial Science, University of Tokyo, Tokyo, Japan*

M. NISHIKAWA • *Renal Regeneration Laboratory, VAGLAHS at Sepulveda & UCLA David Geffen School of Medicine, Los Angels, CA, USA*

MIHO NITOU • *Department of Biology, Faculty of Science, Shizuoka University, Shizuoka, Japan*

TAKAHIRO OCHIYA • *Division of Molecular and Cellular Medicine, National Cancer Center Research Institute, Tokyo, Japan*

YASUYUKI SAKAI • *Institute of Industrial Science, University of Tokyo, Tokyo, Japan*

KHANIT SA-NGIAMSUNTORN • *Department of Pharmacology, Faculty of Medicine Siriraj Hospital, Mahidol University, Bangkok, Thailand*

NOBUYOSHI SHIOJIRI • *Department of Biology, Faculty of Science, Shizuoka University, Shizuoka, Japan*

GOSHI SHIOTA • *Division of Molecular and Genetic Medicine, Department of Genetic Medicine and Regenerative Therapeutics, Graduate School of Medicine, Tottori University, Yonago, Japan*

MAÂMAR SOUIDI • *IRSN, DRPH/SRBE/LTCRA, CEDEX 92262, France*

MINORU TANAKA • *Laboratory of Cell Growth and Differentiation, Institute of Molecular and Cellular Biosciences, The University of Tokyo, Tokyo, Japan*

MERCEDES VAZQUEZ-CHANTADA • *CIC bioGUNE, Technology Park of Bizkaia, Bizkaia, Basque Country, Spain*

XIN WEI WANG • *Laboratory of Human Carcinogenesis, Bethesda, MD, USA*

SHOHEI WAKAO • *Department of Stem Cell Biology and Histology, Tohoku University Graduate School of Medicine, Sendai, Japan*

ADISAK WONGKAJORNSILP • *Department of Pharmacology, Faculty of Medicine Siriraj Hospital, Mahidol University, Bangkok, Thailand*

YOKO YOSHIDA • *Department of Molecular Neuropathology, Tokyo Metroporitan Institute for Neuroscience, Tokyo, Japan*

# Part I

## Several Hepatic Progenitor Cells

# Chapter 1

# Purification and Culture of Fetal Mouse Hepatoblasts that Are Precursors of Mature Hepatocytes and Biliary Epithelial Cells

## Nobuyoshi Shiojiri and Miho Nitou

## Abstract

To investigate cell–cell interactions during mammalian liver development, it is essential to separate hepatoblasts (fetal liver progenitor cells) from nonparenchymal cells, including stellate cells, endothelial cells, and hemopoietic cells. Various factors, which may be produced by nonparenchymal cells, could be assayed for their effects on the growth and maturation of separated hepatoblasts. The protocol using immunomagnetic beads coated with anti-mouse E-cadherin antibody is described for efficient isolation of hepatoblasts from cell suspensions of fetal mouse livers. The purity and recovery rate are larger than 95% and approximately 30%, respectively. The protocol may be useful for various studies focusing on the fetal liver progenitor cells.

**Key words:** Hepatoblasts, Hepatic progenitor cells, E-cadherin, Nonparenchymal cells, Cell–cell interaction, MACS

## 1. Introduction

Hepatoblasts are fetal liver progenitor cells, which have a remarkable growth potential and give rise to both biliary epithelial cells in periportal areas and mature hepatocytes in nonperiportal areas during mammalian development (1, 2). These parenchymal cells are not abundant in the fetal liver that is a hemopoietic organ, in which many hemopoietic cells transiently colonize and proliferate (Fig. 1a, b). Nonparenchymal cells, such as stellate cells, endothelial cells, or hemopoietic cells, control the growth and differentiation of hepatoblasts via several factors, including BMPs, HGF, TNFα, oncostain M, and extracellular matrices that they produce (2).

Takahiro Ochiya (ed.), *Liver Stem Cells: Methods and Protocols*, Methods in Molecular Biology, vol. 826,
DOI 10.1007/978-1-61779-468-1_1, © Springer Science+Business Media, LLC 2012

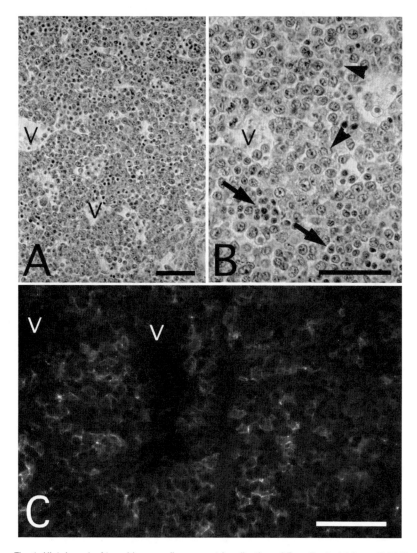

Fig. 1. Histology (**a**, **b**) and immunofluorescent localization of E-cadherin (**c**) in an E 12.5 mouse liver. (**a**, **b**) Hematoxylin–eosin staining. In the liver parenchyma, hepatoblasts with oval nuclei (*arrowheads*) reside among numerous hemopoietic cells (*arrows*). Blood vessels (V) with clear lumina are often observed. (**c**) E-cadherin expression is observed on cell membranes of hepatoblasts, but not detected in other cell types, including endothelial cells, connective tissue cells, hemopoietic cells, and stellate cells. *V* blood vessel. Bars indicate 50 μm.

In order to investigate molecular mechanisms underlying hepatoblast–nonparenchymal cell interactions during hepatic organogenesis, it is indispensable to isolate hepatoblasts from nonparenchymal cells, and to separately culture them. Various factors, which are produced by nonparenchymal cells, could be examined for their effects on the growth and maturation of separated hepatoblasts. Several protocols for isolation of hepatoblasts have been established, including fluorescence-activated cell sorter

(FACS) and magnetic cell sorter (MACS) (3–6). The MACS technique does not require special, expensive equipments or supplies except for antibody-coated magnetic beads. Our isolation protocol for fetal mouse hepatoblasts, which uses anti-E-cadherin antibody-coated magnetic beads, is quite effective also in small laboratories (The purity and recovery rate are larger than 95% and approximately 30%, respectively) (3). Hepatoblasts specifically express E-cadherin, $Ca^{2+}$-dependent epithelial cell adhesion molecule, in the fetal mouse liver whereas nonparenchymal cells do not express this cell adhesion molecule (Fig. 1c) (3).

## 2. Materials

Prepare all solutions using ultrapure water and analytical or cell culture grade reagents under sterile conditions. Store them in a refrigerator before use (unless indicated otherwise). All equipments should be sterilized with autoclaving or heating.

### 2.1. Animals

E12.5 mice (Mice are mated during the night, and noon of the day a vaginal plug found is considered 0.5 days of gestation [E0.5]).

### 2.2. Media, Buffers, and Solutions

(1) 10 mM 2-[4-(hydroxyethyl)-1-piperazinyl]ethanesulfonic acid (HEPES)-buffered Dulbecco's modified Eagle's medium (DMEM) 100 mL.

(2) 10 mM $O,O'$-bis(2-aminoethyl)ethyleneglycol-$N,N,N',N'$-tetraacetic acid (EGTA) dissolved in HEPES-buffered DMEM containing 10% fetal bovine serum (FBS).

(3) 1,000 U/mL dispase (Godo Shusei Co. Ltd., Tokyo, Japan) dissolved in 10% FBS/HEPES-buffered DMEM.

(4) 10% FBS/HEPES-buffered DMEM.

(5) DM-160 (Kyokuto Seiyaku Co. Ltd., Tokyo, Japan) containing 10% FBS and 0.01% deoxyribonuclease I.

(6) 10% FBS/DM-160 100 mL.

(7) 0.1% gelatin in 20 mM tris(hydroxymethyl)aminomethane (Tris)–HCl (pH 7.4)-buffered saline (137 mM NaCl–27 mM KCl; TBS) containing 10 mM $CaCl_2$ and 1% bovine serum albumin (BSA) 100 mL.

(8) Rat anti-mouse E-cadherin antibodies (ECCD-1) (Takara Biomedicals, Otsu, Japan) at 1/1,000 dilution (2 μg/mL) in 1% BSA/TBS.

(9) 1% BSA/TBS 100 mL.

(10) 0.3% Trypan blue in phosphate-buffered saline (PBS).

(11) 10% FBS/DM-160 supplemented with $10^{-7}$ M dexamethasone, penicillin G potassium (100 U/mL), streptomycin sulfate (100 μg/mL) (Culture medium).

The FBS should be heat inactivated for 30 min at 56°C and tested for cytotoxity before use. The medium DM-160 can be replaced by DMEM.

**2.3. Additional Materials Needed**

(1) Dynabeads M-450 sheep anti-rat IgG antibodies (Veritas Corporation, Tokyo, Japan) (The Dynabeads should be prewashed before use according to the procedure in Subheading 3.2.1.).

(2) Dynal Magnetic Particle Concentrator (MPC) (Veritas Corporation).

(3) Plastic centrifuge tubes (15 mL).

(4) Microtubes (1.5 mL).

(5) Nylon mesh filter (132-mm pore size) (Nihon Rikagaku Kikai Co. Ltd., Tokyo, Japan).

(6) Filter holder (13 mm; Millipore Corporation, Billerica, MA, USA).

(7) Scissors.

(8) Watchmaker's forceps.

(9) HT-coated-slides (AR Brown Co. Ltd., Tokyo, Japan).

# 3. Methods

**3.1. Preparation of Cell Suspension from E12.5 Livers**

Carry out all procedures at room temperature unless otherwise specified.

1. Dissect out livers from E12.5 mouse fetuses and then diced in 10 mM HEPES-buffered DMEM in a 30-mm plastic dish with scissors and watchmaker's forceps under a dissection microscope.

2. Subsequently treat diced livers with 10 mM EGTA dissolved in HEPES-buffered DMEM containing 10% FBS on ice for 30 min after removing HEPES-buffered DMEM with a Pasteur pipette. An E12.5 liver requires 500 μL of the EGTA solution for pretreatment. The treatment at step 2 reduces contamination of nonparenchymal cells in a hepatoblast fraction, but the contamination of hepatoblasts increases in a nonparenchymal fraction. The step should be skipped when a nonparenchymal fraction having less contamination of hepatoblasts is required.

3. After removing the EGTA solution, transfer tissues to a 15-mL centrifuge tube, and treat them with 1,000 U/ml dispase dissolved in 10% FBS/HEPES-buffered DMEM for 30 min

at 37°C. An E12.5 liver requires 500 μL of the dispase solution. Pipette the cell suspension gently every 10 min using a Pasteur pipette.

4. After gentle pipetting with a Pasteur pipette, remove undigested tissues from the cell suspension with filtration using a nylon mesh filter (132-mm pore size), which is set in a membrane holder. Recover the cell suspension in a 15-mL centrifuge tube.

5. Collect cells with centrifugation at $60 \times g$ for 10 min at 4°C.

6. Wash the resultant cellular pellet twice with 5 mL (for 5–10 livers) of DM-160 containing 10% FBS and 0.01% deoxyribonuclease I using centrifugation ($60 \times g$, 10 min).

7. Resuspend the resulting cellular pellet in 1 mL of 10% FBS/DM-160 ($10^6$ cells/mL). The viability should be evaluated and be more than 95% by the trypan blue exclusion test.

8. The cell suspensions (Fig. 2a) should be kept on ice before use.

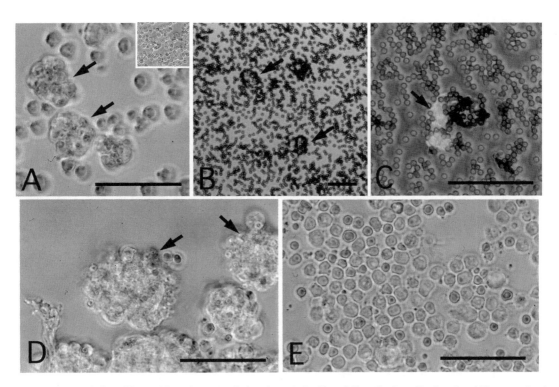

Fig. 2. Immunoisolation of hepatoblasts by magnetic beads coated with anti-E-cadherin antibodies. (**a–e**) Phase-contrast micrographs. (**a**) By dispase digestion, most 12.5-day fetal liver cells are dissociated into single cells (*inset*), but some remain in small clusters (*arrows*). (**b**) After immunomagnetic separation in the presence of $Ca^{2+}$, cells in the hepatoblast fraction form aggregates of different sizes (*arrows*), and single cells are not seen. (**c**) Higher magnification of the hepatoblast fraction. Hepatoblast aggregates are decorated by many beads on their surface (*arrow*). (**d**) Magnetic beads are removed from hepatoblast aggregates (*arrows*) by dispase treatment. (**e**) The majority of the cells in the nonparenchymal cell fraction are single and round. Bars indicate 100 μm.

**3.2. Isolation of Fetal Mouse Hepatoblasts**

The separation of fetal mouse hepatoblasts (3) is based on a modification of the protocol recommended by the bead manufacturer.

*3.2.1. Prewashing of the Dynabeads M-450*

1. Resuspend the Dynabeads M-450 in a vial by gentle vortexing and shaking.

2. Transfer the required amount of Dynabeads M-450 to a washing microtube [12.5 μL ($0.5 \times 10^7$ beads) for five to eight E12.5 livers] using a micropipette. Pipette tips used should be cut off the end.

3. Place the washing tube on a Dynal MPC for 5 min and pipette off the fluid.

4. Remove the microtube from the Dynal MPC and resuspend in an excess volume of washing buffer (500 μL of 1% BSA/TBS).

5. Repeat step 3 and resuspend the washed Dynabead M-450 in washing buffer.

*3.2.2. Cell Separation*

1. Precoat magnetic M-450 Dynabeads covalently coated with sheep anti-mouse IgG antibodies by incubation at 4°C for 60 min with 100 μL of 0.1% gelatin in 20 mM Tris–HCl (pH 7.4)-buffered saline (TBS) containing 10 mM $CaCl_2$ and 1% BSA to prevent nonspecific binding of cells (7).

2. After washing in 1% BSA/TBS, which uses the MPC (Subheading 3.2.1, step 3), add the precoated beads to a solution (500 μL) containing rat monoclonal antibodies against mouse E-cadherin (ECCD-1) at 1/1,000 dilution in 1% BSA/TBS to a final concentration of $10^7$ beads/mL, and incubate for 30 min at room temperature with gentle pipetting every 5 min.

3. Following incubation with ECCD-1, wash the beads twice in 10% FBS/DM-160 using MPC, and add them to 500 μL of the cell suspension ($10^6$ cells/mL) prepared as described above to a final concentration of $10^7$ beads/mL.

4. Incubate the cell suspension in a microtube on ice for 20 min and stir by gentle pipetting every 5 min.

5. Set the microtube in an MPC for 5 min and concentrate hepatoblasts decorated by ECCD-1-coated beads on the tube wall.

6. Transfer the cell suspension, in which hepatoblasts that react with the immunobeads are not contained, to another microtube.

7. Again incubate the cell suspension with fresh magnetic M-450 beads bound with ECCD-1 under the same conditions to recover hepatoblasts that are not reactive in the first separation (Repeat step 4.) (the second separation).

8. Pool the bead fractions obtained at the first and second separation steps (Fig. 2b, c), in which hepatoblasts are contained, and incubate them with 500 μL of dispase (1,000 U/mL) in 10% FBS/DM-160 at 37°C for 40 min to detach beads from separated hepatoblasts.

9. Following the reconcentration of cell-free beads with the MPC, centrifuge the resultant cell suspension at $60 \times g$ for 10 min.

10. Wash twice with centrifugation and resuspend the cellular pellet in 500 μL of 10% FBS/DM-160 supplemented with $10^{-7}$ M dexamethasone, penicillin G potassium (100 U/mL), and streptomycin sulfate (100 μg/mL) (hepatoblast fraction; Fig. 2d).

11. Store the cell suspension, from which hepatoblasts have been removed by treatments with the immunobeads, on ice before use (nonparenchymal cell fraction; Fig. 2e).

12. After centrifugation at $60 \times g$ for 10 min, resuspend nonparenchymal cell fraction in 500 μL of 10% FBS/DM-160 supplemented with $10^{-7}$ M dexamethasone, penicillin G potassium (100 U/mL), and streptomycin sulfate (100 μg/mL).

13. The purity and yield of the hepatoblast and nonparenchymal cell fractions can be checked by RT-PCR analysis for cell type-specific marker genes or immunofluorescent analysis of cytocentrifuge preparations and morphological analysis in short-term culture (24 h) of each fraction (3). Cytocentrifuge preparations can be made by centrifugation of each fraction at $20 \times g$ for 4 min (Cytospin 3, Shandon Scientific Ltd., Cheshire, England). For cytokeratin (hepatoblasts), desmin (stellate cells), PECAM-1 (endothelial cells), and F4/80 (Kupffer cells) immunofluorescence, fix the specimens in acetone at −30°C for 10 min.

14. For culture of separated cell fractions, place the hepatoblast or nonparenchymal cell fraction (70 μL each) on the glass area of HT-coated-slides (AR Brown Co. Ltd., Tokyo, Japan), and incubate at 37°C in a water-saturated atmosphere containing 5% $CO_2$ for 24 or 120 h (3). Also incubate mixtures of both fractions (1:1 v/v), which correspond to a 1:10 mixed population of hepatoblasts and nonparenchymal cells in cell density, under the same conditions after each fraction is concentrated twofold. Add the medium of the conventional culture without separation of hepatoblasts and nonparenchymal cells (at culture hours 48 through 72; conditioned medium [CM]) at 30–50% concentration to cultures of hepatoblast or nonparenchymal cell fractions alone. Separated hepatoblasts have low viability without nonparenchymal cells or the CM (3). Change medium after 24 and 72 h of culture. For immunofluorescence, fix the cultured cells in acetone at −30°C for 10 min after being washed with PBS.

## Acknowledgments

This work was supported in part by Special Coordination Funds for Promoting Science and Technology from the Ministry of Education, Culture, Sports, Science, and Technology, the Japanese Government.

## References

1. Shiojiri, N. (1997) Development and differentiation of bile ducts in the mammalian liver. Microsc Res Tech 39, 328–335.

2. Lemaigre, F. P. (2009) Mechanisms of liver development: concepts for understanding liver disorders and design of novel therapies. Gastroenterology 137, 62–79.

3. Nitou, M., Sugiyama, Y., Ishikawa, K. and Shiojiri, N. (2002) Purification of fetal mouse hepatoblasts by magnetic beads coated with monoclonal anti-E-cadherin antibodies and their in vitro culture. Exp Cell Res 279, 330–343.

4. Yasuchika, K., Hirose, T., Fujii, H., Oe, S., Hasegawa, K., Fujikawa, T., Azuma, H. and Yamaoka, Y. (2002) Establishment of a highly efficient gene transfer system for mouse fetal hepatic progenitor cells. Hepatology 36, 1488–1497.

5. Tanimizu, N., Nishikawa, M., Saito, H., Tsujimura, T. and Miyajima, A. (2003) Isolation of hepatoblasts based on the expression of Dlk/Pref-1. J Cell Sci 116, 1775–1786.

6. Hirose, Y., Itoh, T. and Miyajima, A. (2009) Hedgehog signal activation coordinates proliferation and differentiation of fetal liver progenitor cells. Exp Cell Res 315, 2648–2657.

7. Murphy, S. J., Watt, D. J. and Jones, G. E. (1992) An evaluation of cell separation techniques in a model mixed cell population. J Cell Sci 102, 789–798.

# Chapter 2

## Clinical Uses of Liver Stem Cells

### Yock Young Dan

## Abstract

Liver transplantation offers a definitive cure for many liver and metabolic diseases. However, the complex invasive procedure and paucity of donor liver graft organs limit its clinical applicability. Liver stem cells provide a potentially limitless source of cells that would be useful for a variety of clinical applications. These stem cells or hepatocytes generated from them can be used in cellular transplantation, bioartificial liver devices and drug testing in the development of new drugs. In this chapter, we review the technical aspects of clinical applications of liver stem cells and the progress made to date in the clinical setting. The difficulties and challenges of realizing the potential of these cells are discussed.

**Key words:** Liver stem cells, Liver progenitor cells, Hepatocyte transplantation, Bioartificial liver-assisted device

## 1. Introduction

Liver disease constitutes a major cause of mortality and morbidity worldwide. Despite the fact that the liver has tremendous potential to regenerate, clinical diseases ensue when the regeneration process is exhausted, impaired or is too slow to catch up with the metabolic needs of the liver. Although liver transplantation offers a potential and definitive cure to many patients, the shortage of donor organs, complexity and risks of surgical procedure result in many patients not qualifying for liver transplantation or dying while on the waiting list.

Over the last decade, much effort was spent in exploring the use of hepatocyte transplantation to replace the diseased liver as a simpler and less invasive treatment modality. Buoyed by successful animal experiments showing successful repopulation of diseased liver (1, 2), mature adult human hepatocytes, harvested from donor grafts not used for whole organ transplant, were transplanted

Takahiro Ochiya (ed.), *Liver Stem Cells: Methods and Protocols*, Methods in Molecular Biology, vol. 826,
DOI 10.1007/978-1-61779-468-1_2, © Springer Science+Business Media, LLC 2012

into patients with liver diseases (3). While there were measurable improvements in liver function and correction of metabolic defects, the therapeutic effect is short-lived and limited in terms of overall efficacy and survival. The primary factor limiting the advancement of this approach is the lack of good-quality hepatocytes. This unmet clinical need has thus fuelled the intense search for liver stem cell as a cell source that can generate limitless supply of useful cells for therapeutic purposes.

Liver stem cells would theoretically be able to undergo ex vivo expansion and scale up to produce sufficient numbers that are clinically meaningful for therapeutic purposes (Fig. 1). These stem cells or their differentiated hepatocyte progenies, harvested in environment conforming to good manufacturing practice (GMP) guidelines and free of zoonotic infection risks, can then be cryopreserved in cell banks, allowing immediate access when the cells are needed urgently. They can be transplanted into patients with acute and chronic liver insufficiency, as gene therapy in metabolic diseases, as well as for populating a bioreactor for bioartificial liver dialysis. In addition, mature functional hepatocytes are needed in large amounts in the pharmaceutical industry for toxicology testing

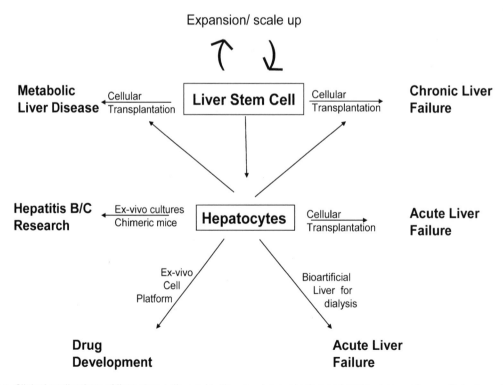

Fig. 1. Clinical applications of liver stem cells would allow ex vivo expansion and scale up to produce sufficient numbers clinically meaningful for therapeutic purposes. They would potentially be useful when transplanted into patients with acute and chronic liver insufficiency, as gene therapy in metabolic diseases, as well as for populating a bioreactor for bioartificial liver dialysis.

in developing new drugs as well as for cellular models studying Hepatitis B and C viruses, screening for new antivirals against these viruses and testing for emergence of resistance.

## 2. Cellular Transplantation

### 2.1. Clinical Evidence (Table 1)

2.1.1. Acute Liver Failure

Acute liver failure presents as an ideal model for cellular transplantation. Massive death of liver cells results in liver insufficiency while the liver scaffold is relatively preserved. If sufficient transplanted hepatocytes can sustain life as a bridge to subsequent recovery of

## Table 1
## Clinical evidence of cellular transplantation

|  | Cells used | Summarized results | References |
|---|---|---|---|
| Fulminant liver failure | Hepatocytes, Fetal hepatocytes, BM MSC | Anecdotal improvement No survival or transplant free benefit | (3, 7) |
| Liver cirrhosis | Hepatocytes BM, MSC, PBSC | Splenic engraftment Improvement in bilirubin/coagulopathy No survival benefit | (10–13) |
| Adjunct treatment | BM CD133+ | Enhanced regeneration of liver lobe | (20) |
| Crigler–Najjar | Hepatocytes | 30–50% Reduction in bilirubin Long-term efficacy | (3, 17) |
| Familial hyperlipidemia | Hepatocytes | 20% Reduction in LDL cholesterol in 3/5 patients | (19) |
| Glycogen storage disease | Hepatocytes | Partial correction | (25) |
| Urea cycle defect | Hepatocytes | Decreased ammonia level No transplant free benefit | (18) |
| Hemophilia | Hepatocytes | Partial correction but still required factor VII | (43) |
| Alpha-1 antitrypsin | Hepatocytes | No benefit | (3) |
| Infantile Refsum's | Hepatocytes | Partial correction | (25) |
| Progressive familial intrahep cholestasis | Hepatocytes | No benefit | (25) |
| Bioartificial liver | Human hepatocytes Porcine hepatocytes HCC stem cells | Improved bilirubin, ammonia, encephalopathy No survival or transplant free benefit | (34) |

BM bone marrow, PBSC peripheral blood stem cell, HCC hepatocellular carcinoma

the native liver or even to transplantation, survival would be improved. The acute nature of this condition requires large number of immediately functional hepatocytes. From a theoretical perspective, stem cells may be too immature to be of immediate use unless the effect is to stimulate regeneration of the native hepatocytes.

Animal models of acute liver failure have demonstrated survival benefit when rescued with primary hepatocytes (4) or hepatocytes derived from stem cells (5, 6). In humans, some two dozen patients with fulminant liver failure have been transplanted to date with adult hepatocytes in an attempt to salvage their failing livers (3). On average, $10^7$–$10^9$ adult hepatocytes were transplanted into the portal circulation for each patient. In contrast, Habibullah et al. (7) transplanted a series of six patients with $10^7$ fetal hepatocytes/kg intraperitoneally. There were transient improvements in encephalopathy and ammonia levels but no overall transplant-free survival benefit was achieved. It is likely that the quantity of cells (up to 5% of liver mass) transplanted for each patient was too low to register a clinical benefit. It was also never established whether these cells did engraft optimally and survive in the hostile liver environment of acute liver failure. Complications were mild but included occasional cases of portal vein thrombosis or sepsis. Using bone marrow stem cells, Gasbarrini et al. (8) transplanted peripheral blood stem cells into a single patient with acute liver failure and showed improvement of liver function over 30 days, although the patient eventually succumbed to sepsis.

*2.1.2. Liver Cirrhosis*

Patients with chronic liver cirrhosis would have a broader therapeutic time window compared to acute liver failure and transplanting stem cells that can expand in vivo and repair the cirrhotic liver would be ideal. In animal studies, both primary hepatocytes (1) and stem cell-derived hepatocytes (9) transplanted into rodent models of chronic liver injury showed improvement in liver functions with single digit percentage in repopulation. Similar transplantation of hepatocytes was performed in humans via the splenic artery with confirmation of engraftment within the splenic pulp. The overall benefit, however, was fairly modest (10). More recently, Terai et al. (11) transplanted bone marrow mesenchymal stem cells (MSCs) into cirrhotic patients via peripheral circulation and showed improvement in liver function, increase in metalloproteinase 9 (MMP-9) activity and increased mitotic frequencies in the liver. Similar improvements in bilirubin and albumin have also been reported by Gordon et al. (12) using peripheral blood stem cells and Lyra et al. (13) using bone marrow stem cell. These studies were not controlled and it is difficult to determine whether the apparent clinical effect is due to transdifferentiation of these stem cells into hepatocytes, remodeling of the cirrhotic liver by MSCs or a regenerative effect from soluble factors coming from the MSCs cultures.

*2.1.3. Metabolic Liver Disease*

Many inborn errors of metabolism can be corrected by liver transplantation (Table 1). As proteins are usually produced in excess in the liver, it was theorized that a small percentage of normal liver cells would be sufficient to correct the clinical defect, making hepatocyte transplantation an attractive treatment modality. Animal models of Crigler–Najjar syndrome (14), tyrosinemia (FAH KO mice) (15) and familial hyperlipidemia (16) were all successfully reversed with transplantation of normal hepatocytes. To date, some 20 patients have been treated with hepatocyte transplantation for a multitude of metabolic diseases (3). The most well-characterized case (17) reported the treatment of a Crigler–Najjar syndrome patient with $1 \times 10^9$ pooled hepatocytes via intraportal transplantation over 15 h. The defect in bilirubin conjugation which causes elevated unconjugated bilirubin was partially corrected and patient was able to come off phototherapy transiently. However, the effect was short lived and clinical disease reverted to baseline after 6 months. Similar experiences were reported in patients transplanted for ornithine transcarbamylase deficiency (18) and familial hypercholesterolemia (19). It is still not known why measurable improvements in metabolic defects after cellular transplantation could not be translated into significant durable clinical improvements. While stem cells with high replicative stimulus may theoretically perform better in such scenarios, there have been no known clinical attempts to date.

*2.1.4. Others*

Liver stem cells may also be useful as adjunct treatment for hepatocellular cancer. Hepatectomy and living donor liver transplantation are frequently limited by small for size thresholds. These limits may potentially be bridged with stem cell adjunct therapy. am Esch et al. (20) transfused CD133+ bone marrow cells in patients undergoing embolization of one lobe of the liver and showed that these CD133+ cells did indeed enhance liver regeneration. With a larger remnant lobe, more aggressive treatment, such as surgery, can be performed on the diseased lobe. In addition, tailored individualized treatment can also be performed with gene therapy for variety of clinical conditions.

# 3. Challenges and Issues with Cellular Transplantation

## 3.1. Cell Source

Using various protocols mimicking physiological development or transdifferentiation induction, hepatocyte-like cells have been derived from multiple stem/progenitor cell sources (Table 2). Pluripotent cells with supposedly infinite replicative potential, such as embryonic stem cells (ESCs) and induced pluripotent stem (iPS) cells have shown efficient generation of functional hepatocytes (21–23). Furthermore, iPS cells do not have the ethical concerns that saddle

## Table 2
## Candidate stem cells: advantages and disadvantages from clinical perspective

| | Ease of derivation | Ease of scale-up | Functional hepatocyte | Oncogenic risk | Rejection risk | Human trials |
|---|---|---|---|---|---|---|
| Hepatocyte | +++ | – | +++ | – | ++ | ++ |
| Fetal hepatocyte | +++ | + | ++ | – | + | ++ |
| Immortalized Hep/FH | ++ | +++ | +++ | ++ | ++ | – |
| ESC | – | + | + | ++ | ++ | – |
| IPSC | – | + | + | ++ | Autologous | – |
| UCSC | + | ++ | + | – | + | – |
| BM/PBSC | +++ | +++ | + | – | ++ | + |
| Adipose SC | ++ | +++ | + | – | Autologous | – |
| HCC cell line | – | +++ | ++ | + | ++ | – |
| Liver stem cell | + | + | ++ | + | ++ | – |
| Amniotic | ++ | ++ | + | – | + | – |
| Wharton's jelly SC | ++ | ++ | + | – | + | – |

*FH* fetal hepatocyte, *ESC* embryonic stem cell, *iPSC* induced pluripotent stem cell, *UCSC* umbilical cord stem cell, *BM* bone marrow, *PBSC* peripheral blood stem cell, *HCC* hepatocellular carcinoma, *SC* stem cell

ESCs and can generate autologous cells for individualized treatment. On the other hand, adult somatic stem cells, such as bone marrow MSCs and adipose tissue stem cells, though less pluripotent, are plentiful in supply and have less oncogenic potential compared to their pluripotent counterparts. Yet from a clinical perspective, the race to the clinical bedside is still on and none of the stem cell candidates are actually out of the gate in having proven their clinical efficacies.

**3.2. Quality of Differentiated Hepatocytes**

The ideal liver stem cell candidate need to demonstrable ability to be scaled-up to clinically meaningful numbers. An attempt to replace 10–20% of human liver mass would require $10^{10}$ cells. Qualitatively, these cells need to be characterized carefully to show purity and uniformity as dedifferentiation, lineage transition and even acquired genetic defects are notorious problems with prolonged culture. Differentiated hepatocytes must show functions that go beyond qualitative assays of albumin or P450 functions. Functions critical for life sustenance, such as detoxification, synthesis of clotting factors, metabolism of ammonia, and excretion of bilirubin, have more direct relevance in the clinical setting. Hepatocytes generated from these stem cell sources need to be comparable in functional magnitude to adult hepatocytes in order for them to have a chance at therapeutic effect. This currently still poses significant challenges to all stem cells at hand and differentiation protocols need to be further optimized.

**3.3. Safety**

Hepatocyte transplantation in humans has established the safety and technical feasibility of the procedure. However, stem cells being pluripotent will always have the theoretical risk of oncogenicity. One undifferentiated ESC is sufficient to form a teratoma when transplanted (24). Uncontrolled replication or epithelial–mesenchymal transition in transplanted stem cells may potentially lead to cancer formation. Adult somatic stem cells though seemingly safer, are not without malignancy risks. Safety tests need to be performed to ensure that cells for transplant have not transformed nor picked up chromosomal anomalies during the scaling-up and differentiation process. These cells would need to be harvested under GMP criteria, with protocols that do not use animal products in the culture process, guaranteeing safety from zoonotic infection. Cryopreservation will also need to be optimized and viability validated before cells are administered to patients.

**3.4. Route of Transplantation**

The technical process of delivering cells to the liver is performed by the introduction of catheter into the portal vein or its tributaries, hepatic artery, or splenic artery (25). The portal vein is usually accessed percutaneously with ultrasound guidance by puncturing through the liver or via a transjugular route from the neck. Arteries are accessed via femoral artery puncture through the groin. The technical challenges include bleeding, especially in acute liver failure patients as clotting function can be severely deranged. Slow infusion of large numbers of cells requires prolonged time as rapid infusion may cause artery or vein thrombosis and raise portal pressure (26). In addition, a catheter may need to be left within portal vasculature if repeated transfusions of large number of cells are required, further increasing the risks of bleeding, infection, and loss of viability of transplanted cells. After infusion into the portal circulation, the transplanted cells move along the portal tracts into the sinusoids and engraft into the hepatic cords by squeezing out between the endothelial lining cells. How efficient this occurs is not really known although kinetic studies in animals have estimated that engraftment is no more than 30% of what is transplanted (27). Animal models have used monocrotaline (28) to injure the endothelial lining cells in the hope that engraftment can be improved, but there are no equivalent maneuvers in human patients.

All described routes of administration, except for intrasplenic puncture, have been used in human patients (29) with measurable improvement in liver functions. However, owing the overall lack of convincing efficacy, it is difficult to determine what is the ideal route of transplantation. Cells transplanted via the splenic artery in animal studies seem to have lower engraftment, and this is believed to be due to the high pressures in the arterial system affecting cell viability (27). In liver cirrhosis, this is even more complicated as portal hypertension may result in reversal of flow (centrifugal flow) in the portal system and cells injected through the portal or splenic veins will not be carried to the liver. This may explain why clinical studies in this scenario have seen large numbers of transplanted cells engrafting in the spleen rather than the liver (10).

Terai et al. (11) transplanted bone marrow MSCs via peripheral venous circulation and reported clinical improvement with increased mitotic activity in the liver. Although homing signals, such as stem cell factor (30), have been reported in animal studies, whether this is sufficient for homing of stem cell to the liver still needs further evaluation. Embolic engraftment of transplanted hepatocytes has been reported in the lungs in animal studies and the implication in the clinical setting is not known (31). Peritoneal injection of hepatocytes, though technically easier, was deemed less ideal as engraftment is random although the use of three-dimensional culture lattices can enhance survival in the peritoneal cavity (25).

### 3.5. Repopulation

With engraftment of only 30% of hepatocytes transplanted, any clinical effect will depend on the successful expansion and significant repopulation of the injured liver. In this aspect, stem or progenitor cells, although lacking function, would have higher replicative potential over time compared to differentiated hepatocytes. The highest repopulation have been seen in animal models with extremely high selection pressure, such as the fumarylacetoacetate hydrolase (FAH) deficient and urokinase-type plasminogen activator (uPA) mice (2, 15). In human transplantation, the transient efficacy of transplantation was likely due to combination of low numbers at engraftment, low replicative pressure in transplanted livers and possible loss through rejection. It is envisaged that preconditioning the liver with some degree of liver injury, including left lobe embolization, partial hepatectomy, chemical injury, or hepatic radiation, may be needed to provide a selective proliferative stimulus to the transplanted cells. Although the idea of deliberately causing more liver injury in patients with liver diseases is worrying, the concept is well entrenched in current protocols of bone marrow transplant, where the bone marrow is preconditioned by high-dose chemotherapy or irradiation before transplant. In the liver, this approach of irradiation preconditioning was proven in principle as reported by Yamanouchi et al. (32). To avoid cell loss through rejection, most protocols have used immunosuppressants similar to that used in whole organ transplantation. In this respect, the use of autologous stem cell sources, such as iPS cells and adult somatic stem cells (adipose tissue or bone marrow MSCs), would obviate this need.

The tougher question is whether the diseased liver is an optimal site for transplanted cells to survive and grow. Livers undergoing acute liver failure and end stage cirrhosis are understandably hostile environments with ongoing avalanche of inflammatory signals, apoptotic pressures and high oxidative stress (33). The notoriously finicky hepatocytes may have difficulty surviving in such an environment and the strategy of cellular transplantation into the liver may be fundamentally flawed. In addition, just as the cirrhotic environment is fertile grounds for hepatocellular carcinogenesis, pluripotent stem cells may be at even higher risk of being transformed.

Current transplantation protocols have focused on replacing only the hepatocyte cell fraction. Increasingly, the niche interaction between parenchymal and nonparenchymal cells is being recognized as critical for stem cell/hepatocyte regeneration and this is not replicated in current strategies of cellular transplantation. A clear understanding of these factors is needed for a more strategic approach to transplantation.

## 4. Bioartificial Liver-Assisted Device

The use of an external bioartificial liver with cells able to provide metabolic functions, detoxify toxins, and synthesize proteins in place of the injured liver would be extremely useful in acute liver failure to bridge patients to recovery or to transplant (34). Current systems incorporate $10^{10}$ cells (150 g or 10% liver weight) in the bioreactor and blood is pumped through the bioreactor in a setup similar to renal dialysis. This approach would bypass the concerns that the liver environment may still be too hostile for engraftment and support the liver until the injury phase is over. Patients with end stage liver disease could also be treated with liver dialysis to prolong and improve the quality of life without having to undergo liver transplantation.

### 4.1. Clinical Evidence

Bioartificial liver device using freshly dissociated hepatocytes from human or porcine livers have been tested in phase I studies. Although improvements can be measured clinically, no definite survival benefits or reduction in requirement for transplant has been proven. One such system currently being evaluated is the extracorporeal liver-assisted device (ELAD) system which incorporates hepatoblastoma-derived HepG2A/C3A cell line (35). Other systems in development include the use of differentiated porcine hepatocytes from PICM-19 cell line (36). This is a pig epiblast derived from inner cell mass at day 8 and is analogous to ES cell lines. In parallel, Kobayashi (37) reported using differentiated hepatocytes from human ES cell KhESC-1 cell line which were capable of producing 351 ng albumin per gram of cells and could metabolize 7.8 and 23.6% of the ammonia and lidocaine loaded in culture. Other interesting approaches used reversible immortalization of adult or fetal hepatocytes (38) by inserting immortalization genes with Cre–lox system. Immortalized hepatocytes can then be expanded into large numbers before the immortalization genes are deleted and functional hepatocytes are incorporated into bioartificial liver device.

**4.2. Issues and Challenges**

Biological cell source is the single most critical factor in bioartificial liver. Like transplantation, such a system would be dependent on the availability of large numbers of functional hepatocytes that can survive and maintain hepatic functions in external bioreactors for significant periods of time. The fact that this is an external system where cells are not transplanted into the human body gives more latitude in the choice of cells as tumorigenicity is less of a concern.

# 5. Drug Testing

Functional hepatocytes with full metabolic profile are needed in large numbers by pharmaceutical industry in the high-throughput screening of new chemical entities as potential new drugs (39). It is estimated that up to 50% of drug withdrawals from the market is related to hepatotoxicity (40). The high cost of investment and low yield begs a high-throughput reliable system capable of screening for hepatotoxicity and also for understanding pharmacokinetics of potential new drugs. The key requirements for such an application are easy availability of cells, consistency in metabolic capacity, and a comprehensive profile of drug metabolizing mechanisms. Currently, pharmaceutical companies use human adult hepatocytes that are not used for transplant or HepaRG (41) cell line derived from hepatocellular carcinoma (HCC). While the former suffers from the same problem of supply and batch to batch variation, HCC cell lines may not have the full drug metabolizing profile of human hepatocytes. Stem cell-derived hepatocytes would be extremely useful for this need. Issues of safety and tumorigenesis are less critical in this setting. iPS cells derived from different patients could even be used in studying phamacogenomics by providing cell platforms that represent the spectrum of population genetic polymorphism. This is particularly useful in defining individualized therapy which may prove to be critical in the treatment of cancer.

# 6. Other Clinical Applications

Stem cell-derived hepatocytes would also be useful as in vitro and in vivo models for testing hepatitis B and C virus (42). These hepatotrophic viruses affect primarily human hepatocytes and the lack of such systems has limited the understanding of pathogenesis of these diseases, development of drugs, and testing for resistance.

## 7. Conclusion

The promise of liver stem cells lies in their ability to continuously generate unlimited supply of stem cells or derived hepatocytes. These cells have shown promise in rescuing animal models of liver injury. Human trials have been less encouraging with progress being limited by cell supply. Current understanding and use of liver stem cells is still in its infancy stage of development. The putative liver stem cell in the adult liver has not been definitively isolated, expanded or even characterized. Derived hepatocytes from ES cells, iPS cells or transdifferentiated cells from adult somatic stem cells have shown principle of proof of efficacy in animal disease models but more work is needed to solve the issues of scaling-up, differentiation efficiency, and functional consistency. At the end of the day, stem cell derived hepatocytes must be comparable to adult hepatocytes in both numbers and quality such that they are clinically useful. Concerns with tumorigenesis especially in pluripotent stem cells continue to be a challenge in cell transplantation and would require long-term safety data before concerns can be put to rest. Much also needs to be done in defining the best way to transplant these cells into patients such that they can regenerate, repair, or replace the injured liver and realize their true potential in saving lives.

## References

1. Kobayashi, N., Ito, M., Nakamura, J., Cai, J., Hammel, J. M., and Fox, I. J. (2000) Treatment of carbon tetrachloride and phenobarbital-induced chronic liver failure with intrasplenic hepatocyte transplantation, *Cell transplantation 9*, 671–673.

2. Rhim, J. A., Sandgren, E. P., Degen, J. L., Palmiter, R. D., and Brinster, R. L. (1994) Replacement of diseased mouse liver by hepatic cell transplantation, *Science 263*, 1149–1152.

3. Strom, S. C., Chowdhury, J. R., and Fox, I. J. (1999) Hepatocyte transplantation for the treatment of human disease, *Semin Liver Dis 19*, 39–48.

4. Ribeiro, J., Nordlinger, B., Ballet, F., Cynober, L., Coudray-Lucas, C., Baudrimont, M., Legendre, C., Delelo, R., and Panis, Y. (1992) Intrasplenic hepatocellular transplantation corrects hepatic encephalopathy in portacaval-shunted rats, *Hepatology (Baltimore, MD) 15*, 12–18.

5. Cho, C. H., Parashurama, N., Park, E. Y., Suganuma, K., Nahmias, Y., Park, J., Tilles, A. W., Berthiaume, F., and Yarmush, M. L. (2007) Homogeneous differentiation of hepatocyte-like cells from embryonic stem cells: applications for the treatment of liver failure, *Faseb J.*

6. Soto-Gutierrez, A., Kobayashi, N., Rivas-Carrillo, J. D., Navarro-Alvarez, N., Zhao, D., Okitsu, T., Noguchi, H., Basma, H., Tabata, Y., Chen, Y., Tanaka, K., Narushima, M., Miki, A., Ueda, T., Jun, H. S., Yoon, J. W., Lebkowski, J., Tanaka, N., and Fox, I. J. (2006) Reversal of mouse hepatic failure using an implanted liver-assist device containing ES cell-derived hepatocytes, *Nat Biotechnol 24*, 1412–1419.

7. Habibullah, C. M., Syed, I. H., Qamar, A., and Taher-Uz, Z. (1994) Human fetal hepatocyte transplantation in patients with fulminant hepatic failure, *Transplantation 58*, 951–952.

8. Gasbarrini, A., Rapaccini, G. L., Rutella, S., Zocco, M. A., Tittoto, P., Leone, G., Pola, P., Gasbarrini, G., and Di Campli, C. (2007) Rescue therapy by portal infusion of autologous stem cells in a case of drug-induced hepatitis, *Dig Liver Dis 39*, 878–882.

9. Yamamoto, H., Quinn, G., Asari, A., Yamanokuchi, H., Teratani, T., Terada, M., and Ochiya, T. (2003) Differentiation of embryonic stem cells into hepatocytes: biological functions and therapeutic application, *Hepatology (Baltimore, MD) 37*, 983–993.

10. Mito, M., Kusano, M., and Kawaura, Y. (1992) Hepatocyte transplantation in man, *Transplant Proc 24*, 3052–3053.

11. Terai, S., Ishikawa, T., Omori, K., Aoyama, K., Marumoto, Y., Urata, Y., Yokoyama, Y., Uchida, K., Yamasaki, T., Fujii, Y., Okita, K., and Sakaida, I. (2006) Improved liver function in patients with liver cirrhosis after autologous bone marrow cell infusion therapy, *Stem cells (Dayton, Ohio) 24*, 2292–2298.

12. Gordon, M. Y., Levicar, N., Pai, M., Bachellier, P., Dimarakis, I., Al-Allaf, F., M'Hamdi, H., Thalji, T., Welsh, J. P., Marley, S. B., Davies, J., Dazzi, F., Marelli-Berg, F., Tait, P., Playford, R., Jiao, L., Jensen, S., Nicholls, J. P., Ayav, A., Nohandani, M., Farzaneh, F., Gaken, J., Dodge, R., Alison, M., Apperley, J. F., Lechler, R., and Habib, N. A. (2006) Characterization and clinical application of human CD34+ stem/progenitor cell populations mobilized into the blood by granulocyte colony-stimulating factor, *Stem cells (Dayton, Ohio) 24*, 1822–1830.

13. Lyra, A. C., Soares, M. B., da Silva, L. F., Fortes, M. F., Silva, A. G., Mota, A. C., Oliveira, S. A., Braga, E. L., de Carvalho, W. A., Genser, B., dos Santos, R. R., and Lyra, L. G. (2007) Feasibility and safety of autologous bone marrow mononuclear cell transplantation in patients with advanced chronic liver disease, *World J Gastroenterol 13*, 1067–1073.

14. Demetriou, A. A., Whiting, J. F., Feldman, D., Levenson, S. M., Chowdhury, N. R., Moscioni, A. D., Kram, M., and Chowdhury, J. R. (1986) Replacement of liver function in rats by transplantation of microcarrier-attached hepatocytes, *Science (New York, NY) 233*, 1190–1192.

15. Overturf, K., al-Dhalimy, M., Ou, C. N., Finegold, M., and Grompe, M. (1997) Serial transplantation reveals the stem-cell-like regenerative potential of adult mouse hepatocytes, *The American journal of pathology 151*, 1273–1280.

16. Eguchi, S., Rozga, J., Lebow, L. T., Chen, S. C., Wang, C. C., Rosenthal, R., Fogli, L., Hewitt, W. R., Middleton, Y., and Demetriou, A. A. (1996) Treatment of hypercholesterolemia in the Watanabe rabbit using allogeneic hepatocellular transplantation under a regeneration stimulus, *Transplantation 62*, 588–593.

17. Fox, I. J., Chowdhury, J. R., Kaufman, S. S., Goertzen, T. C., Chowdhury, N. R., Warkentin, P. I., Dorko, K., Sauter, B. V., and Strom, S. C. (1998) Treatment of the Crigler-Najjar syndrome type I with hepatocyte transplantation, *The New England journal of medicine 338*, 1422–1426.

18. Strom, S. C., Fisher, R. A., Rubinstein, W. S., Barranger, J. A., Towbin, R. B., Charron, M., Mieles, L., Pisarov, L. A., Dorko, K., Thompson, M. T., and Reyes, J. (1997) Transplantation of human hepatocytes, *Transplant Proc 29*, 2103–2106.

19. Grossman, M., Rader, D. J., Muller, D. W., Kolansky, D. M., Kozarsky, K., Clark, B. J., 3rd, Stein, E. A., Lupien, P. J., Brewer, H. B., Jr., Raper, S. E., and et al. (1995) A pilot study of ex vivo gene therapy for homozygous familial hypercholesterolaemia, *Nature medicine 1*, 1148–1154.

20. am Esch, J. S., 2nd, Knoefel, W. T., Klein, M., Ghodsizad, A., Fuerst, G., Poll, L. W., Piechaczek, C., Burchardt, E. R., Feifel, N., Stoldt, V., Stockschlader, M., Stoecklein, N., Tustas, R. Y., Eisenberger, C. F., Peiper, M., Haussinger, D., and Hosch, S. B. (2005) Portal application of autologous CD133+ bone marrow cells to the liver: a novel concept to support hepatic regeneration, *Stem cells (Dayton, Ohio) 23*, 463–470.

21. Basma, H., Soto-Gutierrez, A., Yannam, G. R., Liu, L., Ito, R., Yamamoto, T., Ellis, E., Carson, S. D., Sato, S., Chen, Y., Muirhead, D., Navarro-Alvarez, N., Wong, R. J., Roy-Chowdhury, J., Platt, J. L., Mercer, D. F., Miller, J. D., Strom, S. C., Kobayashi, N., and Fox, I. J. (2009) Differentiation and transplantation of human embryonic stem cell-derived hepatocytes, *Gastroenterology 136*, 990–999.

22. Sullivan, G. J., Hay, D. C., Park, I. H., Fletcher, J., Hannoun, Z., Payne, C. M., Dalgetty, D., Black, J. R., Ross, J. A., Samuel, K., Wang, G., Daley, G. Q., Lee, J. H., Church, G. M., Forbes, S. J., Iredale, J. P., and Wilmut, I. Generation of functional human hepatic endoderm from human induced pluripotent stem cells, *Hepatology 51*, 329–335.

23. Si-Tayeb, K., Noto, F. K., Nagaoka, M., Li, J., Battle, M. A., Duris, C., North, P. E., Dalton, S., and Duncan, S. A. Highly efficient generation of human hepatocyte-like cells from induced pluripotent stem cells, *Hepatology 51*, 297–305.

24. Ishii, T., Yasuchika, K., Machimoto, T., Kamo, N., Komori, J., Konishi, S., Suemori, H., Nakatsuji, N., Saito, M., Kohno, K., Uemoto, S., and Ikai, I. (2007) Transplantation of embryonic stem cell-derived endodermal cells into mice with induced lethal liver damage, *Stem cells (Dayton, Ohio) 25*, 3252–3260.

25. Puppi, J., and Dhawan, A. (2009) Human hepatocyte transplantation overview, *Methods Mol Biol 481*, 1–16.

26. Gagandeep, S., Rajvanshi, P., Sokhi, R. P., Slehria, S., Palestro, C. J., Bhargava, K. K., and Gupta, S. (2000) Transplanted hepatocytes engraft, survive, and proliferate in the liver of

rats with carbon tetrachloride-induced cirrhosis, *J Pathol 191*, 78–85.

27. Chandan Guha, S. S. G., Sung W. Lee, Namita Roy Chowdhury, and Jayanta Roy Chowdhury Hepatocyte Transplantation and Liver-Directed Gene Therapy, Landes Bioscience. Marie Curie Bioscience Database.

28. Witek, R. P., Fisher, S. H., and Petersen, B. E. (2005) Monocrotaline, an alternative to retrorsine-based hepatocyte transplantation in rodents, *Cell transplantation 14*, 41–47.

29. Strom, S. C., Fisher, R. A., Thompson, M. T., Sanyal, A. J., Cole, P. E., Ham, J. M., and Posner, M. P. (1997) Hepatocyte transplantation as a bridge to orthotopic liver transplantation in terminal liver failure, *Transplantation 63*, 559–569.

30. Hatch, H. M., Zheng, D., Jorgensen, M. L., and Petersen, B. E. (2002) SDF-1alpha/CXCR4: a mechanism for hepatic oval cell activation and bone marrow stem cell recruitment to the injured liver of rats, *Cloning Stem Cells 4*, 339–351.

31. Bilir, B. M., Guinette, D., Karrer, F., Kumpe, D. A., Krysl, J., Stephens, J., McGavran, L., Ostrowska, A., and Durham, J. (2000) Hepatocyte transplantation in acute liver failure, *Liver Transpl 6*, 32–40.

32. Yamanouchi, K., Zhou, H., Roy-Chowdhury, N., Macaluso, F., Liu, L., Yamamoto, T., Yannam, G. R., Enke, C., Solberg, T. D., Adelson, A. B., Platt, J. L., Fox, I. J., Roy-Chowdhury, J., and Guha, C. (2009) Hepatic irradiation augments engraftment of donor cells following hepatocyte transplantation, *Hepatology (Baltimore, MD) 49*, 258–267.

33. Friedman, S. L. (2008) Mechanisms of hepatic fibrogenesis, *Gastroenterology 134*, 1655–1669.

34. Streetz, K. L. (2008) Bio-artificial liver devices-tentative, but promising progress, *Journal of hepatology 48*, 189–191.

35. Ellis, A. J., Hughes, R. D., Wendon, J. A., Dunne, J., Langley, P. G., Kelly, J. H., Gislason, G. T., Sussman, N. L., and Williams, R. (1996) Pilot-controlled trial of the extracorporeal liver assist device in acute liver failure, *Hepatology 24*, 1446–1451.

36. Talbot, N. C., Rexroad, C. E., Jr., Powell, A. M., Pursel, V. G., Caperna, T. J., Ogg, S. L., and Nel, N. D. (1994) A continuous culture of pluripotent fetal hepatocytes derived from the 8-day epiblast of the pig, *In Vitro Cell Dev Biol Anim 30A*, 843–850.

37. Kobayashi, N. (2009) Life support of artificial liver: development of a bioartificial liver to treat liver failure, *J Hepatobiliary Pancreat Surg 16*, 113–117.

38. Poyck, P. P., van Wijk, A. C., van der Hoeven, T. V., de Waart, D. R., Chamuleau, R. A., van Gulik, T. M., Oude Elferink, R. P., and Hoekstra, R. (2008) Evaluation of a new immortalized human fetal liver cell line (cBAL111) for application in bioartificial liver, *J Hepatol 48*, 266–275.

39. Bleicher, K. H., Bohm, H. J., Muller, K., and Alanine, A. I. (2003) Hit and lead generation: beyond high-throughput screening, *Nat Rev Drug Discov 2*, 369–378.

40. Lee, W. M. (2003) Acute liver failure in the United States, *Semin Liver Dis 23*, 217–226.

41. Guillouzo, A., Corlu, A., Aninat, C., Glaise, D., Morel, F., and Guguen-Guillouzo, C. (2007) The human hepatoma HepaRG cells: a highly differentiated model for studies of liver metabolism and toxicity of xenobiotics, *Chem Biol Interact 168*, 66–73.

42. Lazaro, C. A., Chang, M., Tang, W., Campbell, J., Sullivan, D. G., Gretch, D. R., Corey, L., Coombs, R. W., and Fausto, N. (2007) Hepatitis C virus replication in transfected and serum-infected cultured human fetal hepatocytes, *The American journal of pathology 170*, 478–489.

43. Dhawan, A., Mitry, R. R., Hughes, R. D., Lehec, S., Terry, C., Bansal, S., Arya, R., Wade, J. J., Verma, A., Heaton, N. D., Rela, M., and Mieli-Vergani, G. (2004) Hepatocyte transplantation for inherited factor VII deficiency, *Transplantation 78*, 1812–1814.

# Chapter 3

# Identification and Isolation of Adult Liver Stem/Progenitor Cells

## Minoru Tanaka and Atsushi Miyajima

## Abstract

Hepatoblasts are considered to be liver stem/progenitor cells in the fetus because they propagate and differentiate into two types of liver epithelial cells, hepatocytes and cholangiocytes. In adults, oval cells that emerge in severely injured liver are considered facultative hepatic stem/progenitor cells. However, the nature of oval cells has remained unclear for long time due to the lack of a method to isolate them. It has also been unclear whether liver stem/progenitor cells exist in normal adult liver. Recently, we and others have successfully identified oval cells and adult liver stem/progenitor cells. Here, we describe the identification and isolation of mouse liver stem/progenitor cells by utilizing antibodies against specific cell surface marker molecules.

**Key words:** Liver stem/progenitor cells, Oval cells, Flow cytometry, Antibody

## 1. Introduction

The adult liver has a remarkable potential to regenerate when injured. In many cases, hepatocytes replicate to repair the damage (1, 2). However, if the injury limits the proliferation of hepatocytes, facultative progenitor cells proliferate around portal veins; this is known as a ductal reaction (3, 4). These proliferating epithelial cells, often referred to as "oval cells" in rodents, are believed to contribute to liver regeneration (5). The nature of oval cells as liver stem cells has been debated based on numerous studies using various rodent models. In mice, a diet containing 3,5-diethoxycarbonyl-1,4-dihydro-collidine (DDC) and a choline-deficient, ethionine-supplemented (CDE) diet have been developed to induce oval cell activation (6–8). Although these proliferating epithelial cells upon

Takahiro Ochiya (ed.), *Liver Stem Cells: Methods and Protocols*, Methods in Molecular Biology, vol. 826,
DOI 10.1007/978-1-61779-468-1_3, © Springer Science+Business Media, LLC 2012

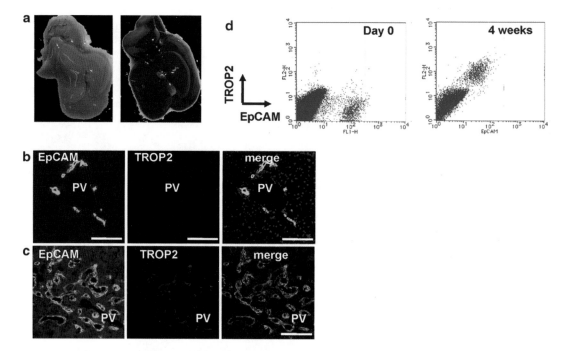

Fig. 1. The DDC-containing diet causes hepatic injury and oval cell activation. (a) Appearance of normal liver (*left panel*) and liver damaged by DDC (*right panel*). (b, c) Immunohistochemistry of frozen sections of normal liver (b) and the liver of mice fed DDC for 5 weeks (c) with anti-EpCAM and anti-TROP2 Abs. TROP2 is expressed in oval cells but not in normal cholangiocytes. (d) Flow cytometric analysis of nonparenchymal cells with anti-EpCAM and anti-TROP2 Abs after DDC feeding. *PV* portal vein. Scale bar, 100 μm. This figure is reproduced from reference (12).

injury by various insults are collectively referred to as oval cells, it remains unclear whether or not those cells in different species by different protocols have common characteristics. To address this issue, oval cells need to be precisely identified and isolated for clonal analysis. Recently, we and others have identified and characterized adult liver stem/progenitor cells by utilizing cell surface markers (9–13). We identified TROP2 and epithelial cell adhesion molecule (EpCAM) as markers expressed on oval cells (Fig. 1) and by using antibodies, isolated the oval cell compartment from the injured livers of mice fed a diet containing DDC. Furthermore, we developed culture system and characterized EpCAM+ cells isolated from normal as well as injured livers. Here, we describe the methods for identifying and isolating mouse liver stem/progenitor cells.

## 2. Materials

### 2.1. Generation of Oval Cells in Mouse Liver

1. Adult male mice (8–12 weeks old). C57BL/6 mice were used for all experiments.
2. Diet containing 0.1% DDC (Bio-Serv).

**2.2. Immuno-histochemistry of Frozen Liver Sections by Zamboni's Fixation (A Modified Paraformaldehyde-Based Fixation Method)**

1. Solution A: Saturated Picric acid solution in distilled water available commercially from Sigma-Aldrich, etc. Any solid materials were removed by filtration before use.

2. Solution B: Twenty grams of paraformaldehyde were dissolved in 100 mL of distilled water. The solution was warmed to 60°C and stirred to melt the paraformaldehyde. A small amount of 1 N NaOH was added if necessary. After complete dissolution, the solution was cooled down. If necessary, it was filtrated.

3. Zamboni's solution: 150 mL of solution A and 100 mL of solution B were mixed well, and 750 mL of PBS (pH 7.4) added to adjust the total volume to 1 L. This solution can be stored at 4°C for at least 1 year.

**2.3. Isolation of EpCAM+ Cells from Livers of Normal and DDC Diet-Fed Mice**

1. Perfusion solution I: Liver Perfusion Medium (GIBCO 17701-038).

2. Perfusion solution II: 8 g NaCl, 0.4 g KCl, 0.56 g $CaCl_2$, 0.078 g $NaH_2PO_4 \cdot 2H_2O$, 0.151 g $Na_2HPO_4 \cdot 12H_2O$, 0.9 g glucose, 0.35 g $NaHCO_3$, and 2.38 g HEPES were dissolved in distilled and deionized water [made up to 1 L and filtered through a 0.22-μm STERICUP (Millipore)].

3. Collagenase solution I: 50 mg of collagenase (Sigma C-5138), 25 mg of DNaseI (Sigma DN25-1G), and 10 mL of FBS were added to 90 mL of perfusion solution II, mixed well, and filtered through a 0.22-μm STERICUP (Millipore) (made fresh as required) (see Note 1).

4. Collagenase solution II: 50 mg of pronase (Roche) was dissolved in 100 mL of Collagenase solution I, mixed well, and filtered through a 0.22-μm STERICUP (Millipore) (made fresh as required).

5. Washing solution: William's E medium containing 10% FBS.

6. Hemolysis buffer: 1 g of Trizma base (Sigma) and 2.8 g of $NH_4Cl$ were dissolved in 500 mL of distilled and deionized water.

7. Catheter of 24 G×3/4" indwelling needle (TERUMO, SR-OT2419C).

8. Peristaltic pump: Masterflex L/S Variable-Speed Modular Drives (HV-07553-80) with silicone tubing (HV-96410-13) (Cole-Parmer, IL).

9. Antibodies: Biotinylated goat anti-TROP2 antibody (BAF1122) (R&D Systems), rat anti-EpCAM antibody (Clone G8.8) (BioLegend) or (Clone 2-17) (MBL International), and FcBlock (BD).

**2.4. Culture of EpCAM+ Cells Isolated from Normal and DDC Diet-Fed Liver**

1. Standard medium: Williams' medium E containing 10% FBS, 10 mM nicotinamide, 2 mM L-glutamine, 0.2 mM ascorbic acid, 20 mM HEPES pH 7.5, 1 mM sodium pyruvate, 17.6 mM $NaHCO_3$, 14 mM glucose, 100 nM dexamethasone,

1× ITS (insulin, transferrin, selenium X) (GIBCO) and 50 µg/mL gentamicin.

2. Cytokines: Human EGF and human recombinant HGF (final concentration of 10 ng/mL each).

3. Type-I collagen-coated dish: 35 mm dishes were coated with Type-I collagen solution (Cellmatrix Type I-C, Nitta gelatin).

4. Trypsin: 0.25% Trypsin–EDTA (#25200)(GIBCO).

# 3. Methods

### 3.1. Generation of Oval Cells and Immuno-histochemistry

1. Mice were fed by the DDC-containing diet for more than 4 weeks (see Note 2). After sacrifice, the whole livers were removed carefully, dissected into pieces of 5 mm and fixed by incubation in Zamboni's solution at 4°C for 8 h to O/N (see Note 3).

2. The fixed tissues were immersed in 10% sucrose/PBS, 15% sucrose/PBS, and 20% sucrose/PBS for 1 day each. Then, the tissues were frozen in OCT compound. Sections (8 µm) were prepared with a cryostat and incubated with each antibody, followed by a fluorescein-conjugated secondary antibody (see Note 4).

### 3.2. Isolation of EpCAM+ Cells from Mouse Liver

1. Liver perfusion solution I (20 mL per a mouse) and collagenase solution I (25 mL per mouse) were warmed at 37°C in a water bath, and a 50-mL glass beaker containing collagenase solution II (10 mL per mouse) was placed in an air incubator at 37°C.

2. An anesthetized mouse was fixed on a cork board and 70% ethanol sprayed on the abdominal skin. The abdomen was opened and the intestine was moved over to the right side (Fig. 2a). The portal vein was cut with surgical scissors without broken away. The bleeding was absorbed with cotton gauze to prevent blood from pooling (Fig. 2b). The hole was cannulated with a catheter connected to a peristaltic pump (Fig. 2c) (see Note 5). Perfusion solution I was delivered at a speed of 3 mL/min (Fig. 2d) and then the vena cava inferior was cut (Fig. 2e). After perfusion for 5 min (total volume, approximately 15 mL) (Fig. 2f), perfusion solution I was replaced with collagenase solution I and the perfusion continued for 8 min (total volume, approximately 25 mL).

3. After digestion, the whole liver was carefully removed and transferred to a Petri dish (10 cm in diameter) (see Note 6). 20 mL of washing solution was added to the dish and the liver capsule was torn by crossing two pairs of fine tweezers. The contents of the liver were dispersed in the solution and

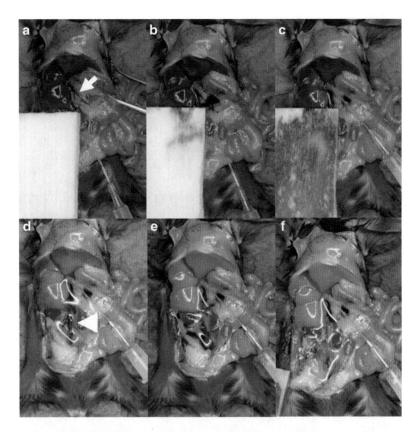

Fig. 2. Perfusion of mouse liver. (**a**) Appearance of abdominal cavity. Arrow indicates portal vein (**b**) Absorption of bleeding with cotton gauze. (**c**) Cannulation of portal vein. (**d**) Perfusion of liver. Arrowhead indicates dilated vena cava inferior. (**e**) Cut-off of vena cava inferior. (**f**) Appearance of perfused liver. This figure is reproduced from reference (12).

pipetted up and down actively until no lumps remained (see Note 7). Then, the cell suspension was passed through a 70-μm filter to remove undigested clots, including the biliary tree and the filtrate was kept on ice (Cell Suspension A). The residual mass on the filter was transferred to prewarmed collagenase solution II and stirred on a magnetic stirrer for 20 min at 37°C in an air incubator. After digestion, it was pipetted up and down and filtered as previously explained (Cell Suspension B).

4. Cell Suspension A and Cell Suspension B were mixed in a 50 mL disposable tube and washing solution was added to make a total volume of 50 mL. The solution was centrifuged at approximately $100 \times g$ for 2 min to remove parenchymal cells (hepatocytes). The supernatant was transferred to another fresh tube and the centrifugation was repeated twice (see Note 8).

5. After centrifugation of the supernatant at $300 \times g$ for 5 min, the pellet was suspended with 7 mL of hemolysis buffer and kept on ice for 3 min. Then, 8 mL of washing solution was added and centrifuged at $300 \times g$ for 5 min. The pellet was resuspended

with 10 mL of 3% FBS/PBS and filtered with 70 μm mesh to remove debris (see Note 9).

6. After centrifugation at $300 \times g$ for 5 min, the pellet was resuspended in 300–500 μL of 3% FBS/PBS. Nonspecific binding was blocked by incubating with FcBlock, and the cell suspension was incubated with fluorescein-conjugated anti-EpCAM antibody (see Note 10). After the staining of dead cells with propidium iodide, EpCAM+ cells were sorted by FACSVantage SE (see Note 11).

### 3.3. Culture of EpCAM+ Cells Isolated from Normal and DDC Diet-Fed Liver

1. EpCAM+ cells isolated by FACSVantage SE were suspended in the Standard Medium and plated at $1 \times 10^4$ cells per Type-I collagen-coated 35-mm dish. Human EGF and human recombinant HGF were added to the culture to a final concentration of 10 ng/mL each.

2. After 9 days of culture, the proliferating cells were trypsinized, washed, and replated onto new culture dishes (Fig. 3) (see Note 12). They continued to grow even after serial passages for more than 6 months, and possessed the potential to differentiate into hepatocytic and cholangiocytic cell lineages under the adequate culture conditions (12).

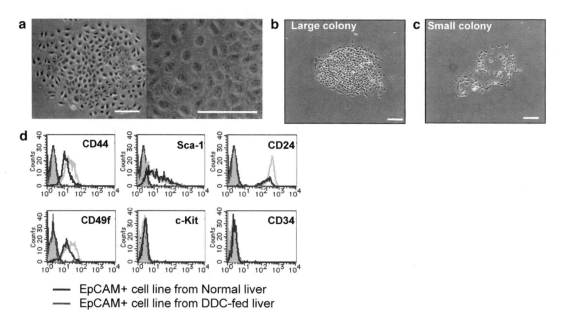

— EpCAM+ cell line from Normal liver
— EpCAM+ cell line from DDC-fed liver

Fig. 3. Characteristics of proliferative EpCAM+ cells. (a) Morphology of proliferative cells with characteristic of liver stem cells derived from EpCAM+ cells of DDC diet-fed livers. (b, c) Appearance of a large colony (b) with the potential to proliferate, and a small colony (c) with limited growth derived from EpCAM+ cells of normal livers. (d) Flow cytometric analysis of some cell surface markers in established cell lines with characteristics of liver stem cells. The cells derived from normal (*blue line*) and DDC diet-fed (*red line*) livers show similar expression profiles of cell surface markers. Scale bar: 200 μm. This figure is reproduced from reference (12).

## 4. Notes

1. Chronic hepatitis induced by the administration of DDC often makes the liver fibrotic. We recommend this collagenase for efficient digestion of the fibrotic liver. Aliquots of collagenase powder were stored at –80°C until used. Avoid freezing and thawing to maintain the enzymatic activity.

2. The formation of oval cells reaches a plateau at about 4 weeks after the administration of DDC. Prolonged administration for more than 8 weeks may cause death.

3. We strongly recommend Zamboni's solution for the detection of TROP2 because the signal is very faint with other reagents, such as 4% PFA and cold acetone.

4. We usually incubated frozen sections with the first antibody solution (1/50–100 dilution with 5% skim milk/PBS) at 4°C for O/N. We used goat anti-TROP2 antibody (BAF1122) (R&D Systems) and rat anti-EpCAM antibody (Clone G8.8) for immunostaining.

5. We usually cannulated with running solution at a very low speed. Put the catheter toward the bleeding portal vein. Do not insert the catheter too deep into the liver or you will damage a vein. If the perfusion is successfully performed, the liver turns yellowish due to blood removal as shown in Fig. 2d.

6. We usually removed the gall bladder carefully without burst because the bile is toxic.

7. We prefer to use a glass pipette with a latex nipple. This permits vigorous pipetting to disperse the clump of tissues rather than electrical pipettes.

8. Repeat this step until the cell pellet is completely dissolved.

9. Insoluble debris should be removed by filtration to proceed to the flow cytometric analysis.

10. We usually use FTIC-conjugated rat anti-EpCAM antibody (Clone 2–17).

11. We recommend a magnetic bead-based cell separation system to enrich EpCAM+ cells before purification by flow cytometry. We usually use the autoMACS system with anti-FITC microbeads before cell sorting by FACSVantage, which saves time for cell sorting and avoids reducing the viability of EpCAM+ cells.

12. The primary culture contains both small and large colonies at this point. The proliferative cells with characteristics of liver stem/progenitor cells are selectively expanded by serial passages.

## References

1. Michalopoulos, G. K. (2010) Liver regeneration after partial hepatectomy: critical analysis of mechanistic dilemmas. *Am J Pathol* **176**, 2–13.

2. Overturf, K., al-Dhalimy, M., Ou, C. N., Finegold, M. and Grompe, M. (1997) Serial transplantation reveals the stem-cell-like regenerative potential of adult mouse hepatocytes. *Am J Pathol* **151**, 1273–1280.

3. Alison, M. R., Golding, M., Sarraf, C. E., Edwards, R. J. and Lalani, E. N. (1996) Liver damage in the rat induces hepatocyte stem cells from biliary epithelial cells. *Gastroenterology* **110**, 1182–1190.

4. Theise, N. D., Saxena, R., Portmann, B. C., Thung, S. N., Yee, H., Chiriboga, L., et al. (1999) The canals of Hering and hepatic stem cells in humans. *Hepatology* **30**, 1425–1433.

5. Farber, E. (1956) Similarities in the sequence of early histological changes induced in the liver of the rat by ethionine, 2-acetylaminofluorene, and 3'-methyl-4-dimethylaminoazobenzene. *Cancer Res* **16**, 142–148.

6. Akhurst, B., Croager, E. J., Farley-Roche, C. A., Ong, J. K., Dumble, M. L., Knight, B., et al. (2001) A modified choline-deficient, ethionine-supplemented diet protocol effectively induces oval cells in mouse liver. *Hepatology* **34**, 519–522.

7. Preisegger, K. H., Factor, V. M., Fuchsbichler, A., Stumptner, C., Denk, H. and Thorgeirsson, S. S. (1999) Atypical ductular proliferation and its inhibition by transforming growth factor beta1 in the 3,5-diethoxycarbonyl-1,4-dihydrocollidine mouse model for chronic alcoholic liver disease. *Lab Invest* **79**, 103–109.

8. Wang, X., Foster, M., Al-Dhalimy, M., Lagasse, E., Finegold, M. and Grompe, M. (2003) The origin and liver repopulating capacity of murine oval cells. *Proc Natl Acad Sci USA* **100** Suppl 1, 11881–11888.

9. Schmelzer, E., Zhang, L., Bruce, A., Wauthier, E., Ludlow, J., Yao, H. L., et al. (2007) Human hepatic stem cells from fetal and postnatal donors. *J Exp Med* **204**, 1973–1987.

10. Yovchev, M. I., Grozdanov, P. N., Zhou, H., Racherla, H., Guha, C. and Dabeva, M. D. (2008) Identification of adult hepatic progenitor cells capable of repopulating injured rat liver. *Hepatology* **47**, 636–647.

11. Suzuki, A., Sekiya, S., Onishi, M., Oshima, N., Kiyonari, H., Nakauchi, H., et al. (2008) Flow cytometric isolation and clonal identification of self-renewing bipotent hepatic progenitor cells in adult mouse liver. *Hepatology* **48**, 1964–1978.

12. Okabe, M., Tsukahara, Y., Tanaka, M., Suzuki, K., Saito, S., Kamiya, Y., et al. (2009) Potential hepatic stem cells reside in EpCAM+ cells of normal and injured mouse liver. *Development* **136**, 1951–1960.

13. Kamiya, A., Kakinuma, S., Yamazaki, Y. and Nakauchi, H. (2009) Enrichment and clonal culture of progenitor cells during mouse postnatal liver development in mice. *Gastroenterology* **137**, 1114–1126, 1126 e1–14.

# Chapter 4

# Isolation and Purification Method of Mouse Fetal Hepatoblasts

## Luc Gailhouste

## Abstract

During development, liver precursors constitute a valuable source of pluripotent stem cells that present the ability to differentiate into both a hepatic and biliary lineage. In the present chapter, we report an experimental procedure developed by our group to isolate mouse fetal hepatoblasts (MFHs) with high purity. The method is based on a selective harvesting of the hepatic parenchymal cells from fetuses (E 14.5), followed by the sorting of E-cadherin+ progenitors through the use of magnetic beads and specific antibodies. This protocol allows the isolation of bipotent liver stem cells expressing both hepatic and biliary markers. Primary cultures of purified MFHs can be maintained under proliferation until confluence, leading to promotion of the differentiation process in the presence of hepatotrophic factors. By using a quantitative real-time polymerase chain reaction approach, we show the hepatospecific phenotype and the progressive maturation of MFHs, delineating early ($\alpha$-fetoprotein), mid (albumin), and late (glucose-6-phosphatase) hepatic markers. Consequently, the model appears to be a valuable cell system for the study of molecular and cellular aspects occurring in hepatic differentiation.

Key words: Mouse fetal hepatoblasts, MFHs, E-cadherin, Cell sorting, Bipotent stem cells, Hepatic differentiation

## 1. Introduction

Effective accessibility to cell resources appears as a major challenge in modern hepatology regarding fundamental research and therapeutic management of liver diseases. An increasing number of reports have focused on the remarkable potential of liver stem cells and their ability to give rise to the hepatic lineage (1). Liver progenitors are known as "oval cells," in reference to their oval-shaped nucleus and scant cytoplasm. These liver-specific stem cells, also called hepatoblasts, exhibit bipotent capacities and are characterized

Takahiro Ochiya (ed.), *Liver Stem Cells: Methods and Protocols*, Methods in Molecular Biology, vol. 826,
DOI 10.1007/978-1-61779-468-1_4, © Springer Science+Business Media, LLC 2012

by an intermediary phenotype between biliary epithelial cells and hepatocytes (2, 3). During embryonic development, hepatoblasts are able to proliferate in order to permit liver morphogenesis and finally differentiate into mature hepatocytes. At that time, the establishment of cell–cell interactions is crucial for organogenesis and implicates a category of glycoprotein named cadherins. In particular, the epithelial cadherin (E-cadherin) represents a key calcium-dependent cell adhesion molecule ensuring the integrity of epithelial tissues (4). In conjunction with hepatotrophic factors, the E-cadherin signaling pathway promotes the onset of a hepatic phenotype and the maturation of embryonic stem cells (5, 6). In contrast to cell differentiation, the alteration of E-cadherin is frequently correlated with cancer progression as reported in numerous human tumors (7). Thus, a reduced expression of the E-cadherin/catenin complex leads generally to the disruption of cell–cell contacts which can promote epithelial to mesenchymal transition, invasiveness and the metastatic potential of a variety of cancers, including hepatocellular carcinoma (8–10).

The molecular and cellular mechanisms leading to the proliferation/differentiation of hepatoblasts are still poorly understood, mainly because of the lack of hepatoblast-like cell lines, as well as the difficulty in isolating liver bipotent progenitors and differentiating primary cultures into mature hepatocytes and/or cholangiocytes. In addition, no technical reports consistently describe the experimental procedures required for the isolation of hepatoblasts and the characterization of their bipotency. Nevertheless, recent studies have focused on a number of markers expressed in hepatic precursors that can be suitable for the specific selection of liver stem cells (11–16). In the present chapter, we report in detail an efficient method for an accurate purification of mouse fetal hepatoblasts (MFHs), based on the sorting of E-cadherin⁺ cells through the use of specific antibodies and magnetic beads. Through controlled stimulation with hepatotrophic growth factors, we characterized isolated E-cadherin⁺ MFHs and demonstrated the suitability of this marker for the specific isolation of hepatoblasts from the fetal liver.

## 2. Materials

### 2.1. Animals

C57BL/6J-Jcl pregnant mice (CLEA Japan, Tokyo, Japan). Embryonic livers are harvested at 14.5 days of gestation (see Note 1).

*2.2. Specific Materials*

1. EasySep magnet. StemCell Technologies (Cat. # 18000).

2. Autoclaved nylon filter with a 60-μm pore size. MILLIPORE (Cat. # NY6000010).

3. Collagen type I-coated dishes (35 mm diameter). IWAKI (Cat. # 4000–010).

4. Other materials required: 8-cm diameter funnel, 10-cm dishes (noncoated), sterilized surgical scissors and forceps, 50-mL round-bottom polypropylene tubes for centrifugation, and 200-mL Erlenmeyer for cell decantation.

*2.3. Reagents*

A complete list of the reagents employed for the isolation of MFHs is provided in Table 1.

## Table 1
## List of reagents

| Description | Source | Cat. number |
| --- | --- | --- |
| *Cell sorting* | | |
| EasySep biotin selection kit (Mouse) | StemCell Technologies | # 18556 |
| Biotin anti-CD324 (E-cadherin) | eBioscience | # 13-3249 |
| *Primary culture of hepatoblasts* | | |
| William's medium E | Gibco | # 12551-032 |
| Fetal bovine serum (Fetalclone) | HyClone Thermo Scientific | # SH30088 |
| L-Glutamine (200 mM) | Gibco | # 25030-081 |
| Penicillin (5,000 U/mL)/streptomycin (5,000 μg/mL) | Gibco | # 15070-063 |
| Insulin from bovine pancreas (50 mg) | Sigma | # I5500 |
| Mouse recombinant oncostatin M (25 μg) | Sigma | # O1637 |
| Human recombinant hepatocyte growth factor (10 μg) | PreproTech | # 100-39 |
| Human recombinant epidermal growth factor (0.2 mg) | Sigma | # E9644 |
| Hydrocortisone 21-hemisuccinate (100 mg) | Sigma | # H2270 |
| Dexamethasone (100 mg) | Sigma | # D2915 |
| *Other reagents* | | |
| Liberase TM Research Grade | Roche | # 05401119001 |
| HEPES buffer solution (1 M) | Gibco | # 15630-080 |
| Sodium chloride (NaCl) | Sigma | # S9888 |
| Potassium chloride (KCl) | Sigma | # P5405 |
| Sodium phosphate ($Na_2HPO_4 \cdot 12H_2O$) | Wako | # 196-02835 |
| Calcium chloride ($CaCl_2 \cdot 2H_2O$) | Sigma | # C7902 |
| D-PBS | Sigma | # D8537 |

### 2.4. Reagents Preparation

#### 2.4.1. Culture Medium

To prepare the hepatoblast basal medium, add the following to a 500-mL final volume of William's E medium: 50 mL of fetal bovine serum (FBS) corresponding to a final concentration of 10%, 5 mL of L-glutamine 200 mM (final concentration: 2 mM), and 5 mL of the antibiotic solution.

To prepare the complete culture medium optimized for primary culture of MFHs, prepare and add the following factors to the basal medium as mentioned below (see Fig. 1):

1. HGF (25 ng/mL): Add 1 mL of sterile water in two vials ($2 \times 10$ µg) after having centrifuged the tubes to obtain a solution with a concentration of 20 µg/mL. Make aliquots and store at –20°C (see Note 2 regarding growth factor stability). 625 µL of the stock solution are required to obtain a final concentration of 25 ng/mL for a 500-mL final volume.

2. EGF (25 ng/mL): After having centrifuged the tube, add 1 mL of sterile water in one vial (0.2 mg) to obtain a 200 µg/mL concentration solution. Use 62.5 µL of the stock solution to obtain a 25 ng/mL concentrated culture medium.

3. OSM (12.5 ng/mL): Centrifuge the vial prior to opening. Reconstitute the lyophilized oncostatin M (25 µg) in 1-mL sterile water to a concentration of 25 µg/mL. Make aliquots and store at –20°C (see Note 2). Use 250 µL to get appropriate working concentration in 500 mL of medium.

Prepare stock solutions of the following factors mentioned below:

1. Insulin stock solution (5 mg/mL): Reconstitute the 50 mg of lyophilized powder in a 50-mL tube by adding 10 mL of sterile

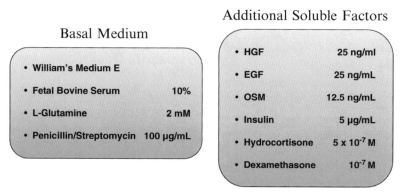

Fig. 1. Complete medium used for the primary culture of mouse fetal hepatoblasts.

water. After dissolution of the insulin, the solution becomes whitish. To obtain a perfectly limpid solution, add one or two drops of 1 N HCl (see Note 3). Perform filtration of the solution with a 0.22-μm filter. Make aliquots and store at –20°C (see Note 4).

2. Hydrocortisone-21-hemissuccinate and dexamethasone stock solutions ($10^{-2}$ M): Add appropriate volumes of sterile water to reconstitute both corticoids at $10^{-2}$ M. Make aliquots and store at –20°C after filtration of the solutions (see Note 4).

Finally, to obtain the complete medium used for hepatoblast differentiation into mature hepatocytes add, respectively: 500 μL of insulin (final concentration: 5 μg/mL), 25 μL of hydrocortisone ($5 \times 10^{-7}$ M), and 5 μL of dexamethasone ($10^{-7}$ M). Keep the complete medium at 4°C (see Note 5).

*2.4.2. HEPES Buffer Solution*

Prepare the HEPES buffer by adding the following compounds to 750 mL of autoclaved distilled water: 8 g NaCl, 0.2 g KCl, 0.1 g $Na_2HPO_4 \cdot 12H_2O$, and 10 mL of HEPES commercial solution (equivalent to 2.38 g). Then, mix well using a magnetic rotating system; adjust volume at 1 L and pH at 7.65. Finally, store HEPES buffer at 4°C after filtration of the solution with a 0.22-μm filter.

*2.4.3. Liver Dissociation Solution*

1. Reconstitute the lyophilized liberase (Roche, Cat. # 05401119001, collagenase amount: 5 mg) in 2 mL of sterile water to get a solution with an enzyme activity equal to 13 U/mL. Place the vial on ice for 30 min to rehydrate the enzyme and gently shake the solution every few minutes until complete dissolution. Immediately store the unused stock solution in single-use aliquots at –20°C (see Note 6 regarding important remarks about liberase storage and stability).

2. We use a working concentration of liberase equivalent to 0.13 U/mL. To prepare 10 mL of dissociation buffer (sufficient for the digestion of approximately 15–20 fetal livers), 100 μL of the stock solution are required. Ten minutes before performing liver dissociation, dilute the concentrated liberase into the HEPES solution containing calcium (see Note 7). To prepare 100 mL of calcium–HEPES buffer, dissolve 56.25 mg of $CaCl_2 \cdot H_2O$. Store at 4°C after 0.22-μm filtration.

*2.4.4. E-Cadherin Cell Sorting Medium*

The recommended medium for the isolation of E-cadherin positive cells is a calcium–HEPES buffer solution supplemented with 2% FBS (see Note 8).

## 3. Method

The following experimental protocol describes the usual method used to isolate MFHs from fetuses (E 14.5) of pregnant mice. The purification of hepatic progenitors relies on two consecutive steps (1) isolation of parenchymal cells after an enzymatic dissociation of the fetal livers; (2) sorting of isolated cells by using the E-cadherin surface marker (see Fig. 2).

### 3.1. Collection of Mouse Fetal Livers

1. Before starting the harvesting procedure, place 10 mL of calcium-supplemented HEPES buffer into a warming water-bath (37°C). Throughout dissection, use cold buffers (4°C)

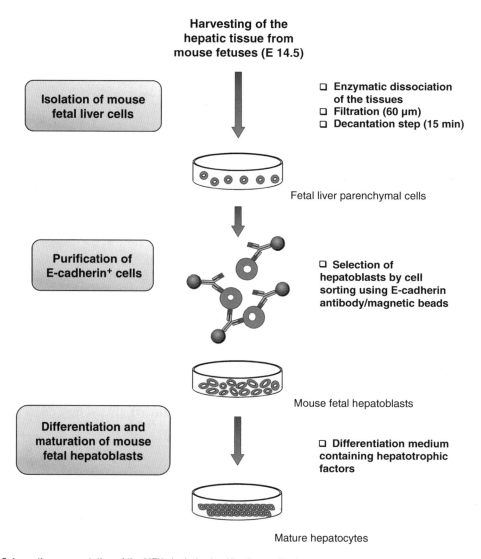

Fig. 2. Schematic representation of the MFHs isolation/purification method.

and keep the fetal livers in HEPES solution on ice. Prepare in advance all the required material as mentioned above.

2. After deep anesthesia, two pregnant mice (E 14.5) are sacrificed by cervical dislocation. Following opening of the abdominal cavity with surgical scissors, localize uterus containing the fetuses. Remove uterus carefully to avoid injury to the fetuses and store it in PBS on ice.

3. Extract each fetus with sterile thin forceps and transfer them to a 10-cm Petri dish containing cold HEPES buffer. Typically, 15–20 fetuses can be collected from two pregnant mice.

4. Using a stereo microscope system, dissect each fetus to isolate the liver, taking care to systematically remove all the surrounding/adjacent tissues that could contaminate the culture (see Note 9). Recognizing the fetal liver is simple due to its strong dark/red coloration (important presence of blood) compared to the rest of the fetus (white).

5. Next, transfer the livers into another dish containing clear cold HEPES and meticulously peel the hepatic capsule.

6. Finally, keep the harvested tissues in HEPES on ice until the end of the collection procedure.

**3.2. Isolation of Parenchymal Hepatic Cells from Fetal Liver**

1. Before starting the isolation protocol, add 100 μL of liberase enzyme (working aliquot) to the 10 mL of sterile warmed calcium–HEPES solution in order to make the dissociation buffer. Keep at 37°C. From this point, all experimental procedures have to be carried out under sterile conditions.

2. After transferring the collected livers into a 1.5-mL tube without medium, hash the samples thoroughly with thin and sterile surgical scissors.

3. Perform enzymatic dissociation as follows: Suspend tissue fragments in 10 mL of the liberase/calcium/HEPES digestion buffer (use a 50-mL tube) and incubate at 37°C for 5 min. Shake the tube gently every minute. Next, a physical dissociation step is required by softly pipetting the lysate several times using a 2-mL serological pipette. The final step to achieve the dissociation of liver fragments is an additional digestion for 5 min at 37°C (see Note 10). From this point onward, absolutely avoid any bubbles that could damage the isolated cells.

4. Add 25 mL of warmed basal medium (without growth factors) to dilute the enzyme and stop the digestion procedure.

5. Perform one filtration of the cell suspension through a 60-μm nylon filter. Wash the filter thoroughly with the basal medium to reach a final approximate volume of 100 mL.

6. Carry out one decantation by simply reserving the cell suspension in a 200-mL Erlenmeyer for 15 min at room temperature (see Note 11 for further information about this essential point).

7. Carefully discard the upper phase (approximately 3/4 of the total volume) with a 25-mL serological pipette.

8. Resuspend cells contained in the 20-mL remaining medium after adding up to 50 mL of the basal medium.

9. Distribute the suspension into two 50-mL round-bottom polypropylene tubes using a 25-mL pipette, afterward adjusting each tube to an equal volume of 40 mL.

10. Centrifuge the tubes at 1,000 rpm ($22 \times g$) for 2 min at 4°C.

11. Remove the supernatants and add $2 \times 250$ μL of HEPES buffer plus calcium and 2% FBS on cells. A moderate pipetting using a 2-mL pipette is necessary to completely dissociate the cell pellet (see Note 12). Pool the suspensions contained in both tubes and keep in a single 50-mL round-bottom polypropylene tube at room temperature.

12. Count cells using a numeration system, typically an improved Neubauer hemocytometer, and determine the cell density (see Note 13). Use trypan blue to assess the viability of isolated cells (see Note 14).

### 3.3. Purification of E-Cadherin Positive MFHs

The number of gathered cells after the dissociation/decantation phase can greatly vary regarding the number of fetuses collected and/or the experimental conditions applied during the liberase treatment. To achieve an effective cell sorting, cell density needs to be precisely adjusted. Here, we describe the method used for processing a cell suspension at the concentration of $10^7$ cells for 100 μL for a total volume between 0.5 and 2.5 mL ($5–25 \times 10^7$ cells) (see Fig. 3).

1. First, prepare and place the cells in a 5-mL round-bottom polystyrene tube ($12 \times 75$ mm) after performing the appropriate dilution in the cell sorting medium (HEPES–calcium–2% FBS).

2. Add 1 μL/$10^7$ cells of the mouse blocking antibody provided with the kit to the suspension and mix wel' by tapping the tube (see Note 15).

3. Add the biotinylated primary antibody (anti-E-cadherin) at a final concentration of 10–20 μg/mL (2–4 μL/$10^7$ cells/100 μL, Biotin anti-CD324, eBioscience) and mix carefully (see Note 16). Incubate at room temperature for 15 min. Tap the tube every few minutes to put sedimented cells into suspension.

4. Put into the cell suspension the biotin selection cocktail at 10 μL/$10^7$ cells. Keep 15 min at room temperature and sometimes homogenize.

5. Add 5 μL/$10^7$ cells of magnetic nanoparticles, mix well, and incubate at room temperature for 10 min taking care to resuspend the cell pellet when necessary. Ensure the homogeny of the

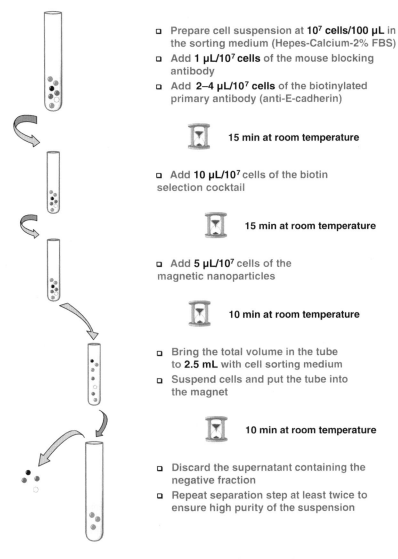

- Prepare cell suspension at **10$^7$ cells/100 μL** in the sorting medium (Hepes-Calcium-2% FBS)
- Add **1 μL/10$^7$ cells** of the mouse blocking antibody
- Add **2–4 μL/10$^7$ cells** of the biotinylated primary antibody (anti-E-cadherin)

**15 min at room temperature**

- Add **10 μL/10$^7$** cells of the biotin selection cocktail

**15 min at room temperature**

- Add **5 μL/10$^7$** cells of the magnetic nanoparticles

**10 min at room temperature**

- Bring the total volume in the tube to **2.5 mL** with cell sorting medium
- Suspend cells and put the tube into the magnet

**10 min at room temperature**

- Discard the supernatant containing the negative fraction
- Repeat separation step at least twice to ensure high purity of the suspension

Mouse fetal hepatoblasts (E-cadherin$^+$)

Fig. 3. Experimental protocol for MFHs sorting.

beads by pipetting the solution vigorously and several times before use.

6. Bring the total volume of the cell suspension to 2.5 mL by adding an adequate volume of cell sorting medium.

7. Resuspend cells by pipetting up and down three times with a 2.5-mL serological pipette. Immediately, put the uncapped tube into the magnet and keep at room temperature for 10 min.

8. Discard the negative fraction by inverting the magnet and the tube for 3 s (see Note 17). Keep the negative fraction if needed.

9. Remove the tube from the magnet and immediately add 2.5 mL of HEPES/calcium/2% FBS medium. Suspend cells by gently pipetting as mentioned above and place the tube back into the magnet for 10 min.

10. Repeat steps 7–9 at least once to ensure effective cell purification (see Note 18 for further information relating to purity).

11. Resuspend the positive selected cells in 500 µL of complete medium containing the hepatic factors, transfer the enriched sample into a new tube and determine the cellular density and viability.

**3.4. Differentiation and Maturation of MFHs**

The E-cadherin+ cells are seeded on collagen type I-coated dishes with a 35-mm diameter at a density of $2 \times 10^5$ cells/cm². 24 h after plating, the attached cells are washed twice with the basal medium (devoid of growth factors). Then, the basal medium is replaced by 2 mL of the complete medium in each dish. As described in the previous section, the medium employed for primary culture is based on a mixture of William's E medium, L-glutamine, and 10% of FBS, supplemented with appropriate factors promoting hepatic differentiation that include HGF, EGF, and OSM. During the culture period, the cells are incubated at 37°C in a humid atmosphere containing 5% $CO_2$. The medium is changed entirely every each days. After cell attachment, MFHs do not require FBS and EGF anymore, and these growth factors can be removed from the medium after 48 h of culture. Following plating on collagen type I, MFHs proliferate to rapidly reach confluence after 3 or 4 days of culture and adopt a compact morphology exhibiting a significant cuboidal appearance (Fig. 4a). In the presence of hepatic factors, undifferentiated cells undergo relative changes of phenotype which include the reduction of growth and the appearance of an oval cell-like shape characterized by a large elliptical nucleus and a slight cytoplasm (Fig. 4b). Finally, oval cells form pronounced cell aggregates and adopt the typical morphology of mature hepatocytes with a small round nucleus and a dark cytoplasm (Fig. 4c, d).

**3.5. Characterization of MFHs**

Hepatotrophic factors and cellular density both appear to condition the terminal differentiation of MFHs *in vitro*. By using a defined hepatospecific medium, we demonstrated that isolated E-cadherin+ liver fetal cells express increased levels of albumin, whereas α-fetoprotein rates decline (Fig. 5). The profiles displayed by both markers are specific to the transition scheme observed in the case of liver regeneration or when hepatic progenitors differentiate into mature hepatocytes during development. Furthermore, the expression of glucose-6-phosphatase which characterizes typically adult mature hepatocytes is strongly enhanced during the induced-differentiation process of purified MFHs (Fig. 6). Notably, MFHs are also characterized by a concomitant expression of α-fetoprotein and cytokeratin 19 (CK19) at the early stage of culture. CK markers are commonly

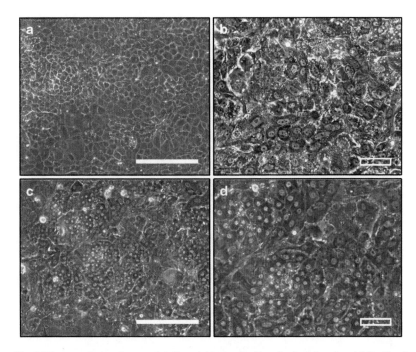

Fig. 4. Primary cultures of mouse fetal hepatoblasts after E-cadherin cell sorting. (**a**) Confluent monolayer formed by the hepatic progenitors 3 days after seeding. (**b**) In the presence of hepatotrophic factors, MFHs undergo radical changes of phenotype from day 5 and adopt an oval cell-like morphology. (**c**, **d**) Mature hepatocytes derived from E-cadherin[+] MFHs after 12 days of culture. *Scale bars* are respectively 200 μm (**a**, **c**) and 50 μm (**b**, **d**).

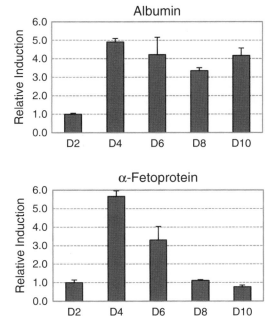

Fig. 5. Expression levels of albumin and α-fetoprotein during MFHs maturation. E-cadherin[+] MFHs were isolated from E 14.5 fetal livers after performing a purification procedure. Total RNAs were harvested after cell seeding at the indicated times. cDNA were synthesized from 1 μg of total RNA of each fraction. After an initial denaturation at 95°C for 2 min, the thermal cycles of quantitative real-time PCR were repeated 40 times as follows: 95°C for 15 s, 60°C for 30 s. The housekeeping gene GAPDH was used as an internal control to normalize the amount of cDNA.

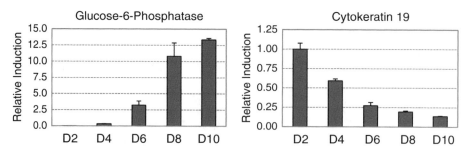

Fig. 6. Expression levels of the hepatospecific enzyme glucose-6-phopshatase and the biliary epithelial cell marker cytokeratin 19 (CK19) during the maturation process of MFHs.

associated with epithelial cells of the biliary tract, such as cholangi-ocytes. This point argues for the bimodal ability of MFHs to differentiate into the hepatic and the biliary lineage. To summarize, the present data based on the expression profiles of the major hepatic markers show that E-cadherin cell sorting from the embry-onic liver is a valuable method for making the isolation of bipotent hepatoblasts possible. MFHs are able to proliferate and are also responsive to growth factor stimulation leading to the induction of hepatic maturation.

## 4. Notes

1. Animals were maintained in an isolator unit under a constant temperature of 20°C and a 12-h light–dark cycle. Mice received standard sterilized food and water ad libitum. All experiments were carried out in accordance with national laws and institu-tional regulations.

2. Reconstituted HGF, EGF, and OSM are stable at least 3 months when stored at –20°C and up to 15 days at 4°C. Avoid repeat-ing freezing/thawing cycles by making appropriate aliquots.

3. The bovine insulin requires acidic conditions in order to be dissolved.

4. After reconstitution, working aliquots of insulin, dexametha-sone, and hydrocortisone can be stored at 4°C up to 3 weeks as well as several months at –20°C. Repeated freezing and thawing is not recommended.

5. The complete medium for hepatoblast primary culture must be used within 2 weeks of its preparation. To avoid degradation of growth factors, warm at 37°C only the volume required for each experiment (cell seeding or medium change).

6. The reconstituted liberase is stable at –20°C for up to 3 months. To avoid repeated freezing and thawing, highly damageable for the enzyme, make stock solution in single-use aliquots.

Furthermore, prepare the liberase–HEPES solution just before performing liver dissociation to avoid enzyme degradation.

7. Calcium is necessary to insure the enzymatic activity of liberase. This cation acts as a cofactor and stabilizes the enzyme. In addition, protease inhibitors, serum, and bovine albumin inhibit the performance of liberase and must be excluded from the dissociation medium.

8. E-cadherin is a transmembrane glycoprotein exhibiting a basic structure in its extracellular portion that is composed of repeating domains, each with two consensus $Ca^{2+}$-binding motifs. Consequently, the addition of calcium in the cell sorting medium ensures E-cadherin integrity.

9. Tissues surrounding the fetal liver represent a major cause of contamination by fibroblasts in primary cultures, making it crucial to perform the most complete dissection of each liver under a microscope system to remove contaminants (portal vein, intestines, etc.).

10. The duration of the enzymatic digestion can vary slightly depending on the aspect of the solution containing tissue fragments. As a rule, the dissociation medium must become cloudy and brown at the end of the procedure traducing the effective dissociation of the hepatic parenchyma. If not, increase the time of enzymatic treatment or use a fresh liberase solution.

11. The decantation procedure is essential to remove from the cell suspension all hematopoietic and nonparenchymal cells. Indeed, the fetal liver is a transient niche of hematopoiesis during the embryonic development and produces massive amounts of blood cells. The decantation step enables a soft sedimentation of the hepatoblasts which typically aggregate to form heavy clusters descending to the bottom. On the contrary, single blood cells and nonparenchymal cells stay in the upper phase of the suspension and can easily be discarded.

12. As mentioned above, hepatic progenitors tend to form aggregates in the presence of divalent cations such as $Ca^{2+}$ that is contained in the medium or HEPES buffer. In consequence, a soft physical dissociation phase is necessary. However, avoid the use of 1-mL tips that could damage the cells, and opt for an automatic system with a 2-mL serological pipette.

13. To prevent the formation of clusters that occurs after tissue dissociation. To facilitate cell counting, use trypsin solution (0.05% + EDTA) when diluting the cell suspension in order to prevent the formation of clusters that occurs after tissue dissociation.

14. Several factors can condition the viability of hepatic progenitors after the isolation and the purification stages. However, the number of viable cells observed remains suitable when following the standard procedure described previously. See Subheading 5 for further details regarding cell viability.

15. It is essential during the procedure to limit cell clumping and sedimentation that could decrease the purification/recovery rate. To ensure cell suspension, tapping the tube is highly preferable to several pipetting steps.

16. In order to achieve an optimal purity and recovery, it is necessary to perform antibody titration prior to carrying out cell sorting. The number of selected cells may be higher with an increased amount of antibody. However, an antibody excess may reduce the purity.

17. During removal of the negative fraction and during washing steps, it is recommended not to shake or blot off any drops that might form at the top of the tube.

18. At least three separation stages into the magnet are essential to isolate E-cadherin-positive cells with high purity. Additional separation rounds may improve the purity. However, it will decrease recovery.

## 5. Important Remarks

1. Purified collagenase that compounds liberase enzyme requires calcium. The exposure of liberase to divalent cation chelators like EDTA removes calcium leading to the inactivation of the enzyme.

2. Numerous critical points can contribute to reducing the viability of isolated cells. In order to obtain a higher viable cell rate (1) limit ischemia by performing the isolation/purification protocol in the briefest time; (2) absolutely avoid the formation of bubbles during pipetting; (3) employ the use of 2-mL serological pipettes rather than 1-mL tips; and (4) adhere to the concentration of liberase and the incubation conditions during the enzymatic dissociation of the fetal livers.

3. Cellular density is a crucial point conditioning hepatic differentiation of liver progenitors. During seeding, take care to ensure optimal homogeny of plated cells in order to avoid high density areas that correlate to a quick over-confluence, as well as low density zones leading to a significant delay in the differentiation and the maturation process of hepatoblasts.

## Acknowledgments

The author thanks Ayako Inoue for her efficient technical assistance, Takahiro Ochiya for his constructive comments on the study, and Takahashi Ryou-u for his advice regarding the cell sorting procedure.

This work was supported by a Grant-in-Aid from the Third-Term Comprehensive 10-Year Strategy for Cancer Control from the Ministry of Health, Labour, Welfare of Japan (H21-001).

## References

1. Ochiya, T., Yamamoto, Y., and Banas, A. (2010) Commitment of stem cells into functional hepatocytes. *Differentiation* 79, 65–73.

2. Fausto, N. (2004) Liver regeneration and repair: hepatocytes, progenitor cells, and stem cells. *Hepatology* 39, 1477–1487.

3. Roskams, T. (2006) Different types of liver progenitor cells and their niches. *J Hepatol* 45, 1–4.

4. Butz, S., and Larue, L. (1995) Expression of catenins during mouse embryonic development and in adult tissues. *Cell Adhes Commun* 3, 337–352.

5. Dasgupta, A., Hughey, R., Lancin, P., Larue, L., and Moghe, P. V. (2005) E-cadherin synergistically induces hepatospecific phenotype and maturation of embryonic stem cells in conjunction with hepatotrophic factors. *Biotechnol Bioeng* 92, 257–266.

6. Moore, R. N., Dasgupta, A., Rajaei, N., Yarmush, M. L., Toner, M., Larue, L., and Moghe, P. V. (2008) Enhanced differentiation of embryonic stem cells using co-cultivation with hepatocytes. . *Biotechnol Bioeng* 101, 1332–1343.

7. Wijnhoven, B. P., Dinjens, W. N., and Pignatelli, M. (2000) E-cadherin-catenin cell-cell adhesion complex and human cancer. *Br J Surg* 87, 992–1005.

8. Berx, G., Cleton-Jansen, A. M., Strumane, K., de Leeuw, W. J., Nollet, F., van Roy, F., and Cornelisse, C. (1996) E-cadherin is inactivated in a majority of invasive human lobular breast cancers by truncation mutations throughout its extracellular domain. *Oncogene* 13, 1919–1925.

9. Wei, Y., Van Nhieu, J. T., Prigent, S., Srivatanakul, P., Tiollais, P., and Buendia, M. A. (2002) Altered expression of E-cadherin in hepatocellular carcinoma: correlations with genetic alterations, beta-catenin expression, and clinical features. *Hepatology* 36, 692–701.

10. Zhai, B., Yan, H. X., Liu, S. Q., Chen, L., Wu, M. C., and Wang, H. Y. (2008) Reduced expression of E-cadherin/catenin complex in hepatocellular carcinomas. *World J Gastroenterol* 14, 5665–5673.

11. Nitou, M., Sugiyama, Y., Ishikawa, K., and Shiojiri, N. (2002) Purification of fetal mouse hepatoblasts by magnetic beads coated with monoclonal anti-e-cadherin antibodies and their in vitro culture. *Exp Cell Res* 279, 330–343.

12. Tanimizu, N., Nishikawa, M., Saito, H., Tsujimura, T., and Miyajima, A. (2003) Isolation of hepatoblasts based on the expression of Dlk/Pref-1. *J Cell Sci* 116, 1775–1786.

13. Dan, Y. Y., Riehle, K. J., Lazaro, C., Teoh, N., Haque, J., Campbell, J. S., and Fausto, N. (2006) Isolation of multipotent progenitor cells from human fetal liver capable of differentiating into liver and mesenchymal lineages. *Proc Proc Natl Acad Sci USA* 103, 9912–9917.

14. Miki, R., Tatsumi, N., Matsumoto, K., and Yokouchi, Y. (2008) New primary culture systems to study the differentiation and proliferation of mouse fetal hepatoblasts. *Am J Physiol Gastrointest Liver Physiol* 294, 529–539.

15. Oertel, M., Menthena, A., Chen, Y. Q., Teisner, B., Jensen, C. H., and Shafritz, D. A. (2008) Purification of fetal liver stem/progenitor cells containing all the repopulation potential for normal adult rat liver. *Gastroenterology* 134, 823–832.

16. Okabe, M., Tsukahara, Y., Tanaka, M., Suzuki, K., Saito, S., Kamiya, Y., Tsujimura, T., Nakamura, K., and Miyajima, A. (2009) Potential hepatic stem cells reside in EpCAM + cells of normal and injured mouse liver. *Development* 136, 1951–1960.

# Chapter 5

# Isolation of Hepatic Progenitor Cells from the Galactosamine-Treated Rat Liver

## Norihisa Ichinohe, Junko Kon, and Toshihiro Mitaka

## Abstract

Oval cells and small hepatocytes (SHs) are well known as hepatic stem/progenitor cells. However, the relationship between the oval cells and SHs in liver regeneration is not well understood. To resolve this issue, we established a technique to selectively separate oval cells and SHs. In the injured rat liver, oval cells and SHs transiently appear in the initial period of liver regeneration. Thy1[+] and CD44[+] cells are candidates for markers of oval cells and SHs, respectively. In this chapter, the methods for sorting and culture of the cells are described in detail.

**Key words:** Oval cells, Small hepatocytes, Thy1, CD44, Liver, Stem/progenitor cells, Galactosamine

## 1. Introduction

It is well known that hepatic stem/progenitor cells are activated when the proliferation of mature hepatocytes (MHs) is inhibited by hepatotoxins (1–6). Of these hepatic stem/progenitor cells, oval cells and small hepatocytes (SHs) are well recognized. Oval cells (7), named for their possession of ovoid nuclei, are known to express markers for biliary epithelial cells, e.g., cytokeratin (CK) 7 and CK19, and for hepatoblasts, e.g., AFP and cell membrane proteins such as CD34, c-kit, and Thy-1, shared hematopoietic stem cell markers (8). On the contrary, SHs are a subpopulation of hepatocytes (9). Their size is less than half that of MHs and they possess hepatic characteristics. These cells can clonally proliferate and mature by interacting with hepatic nonparenchymal cells (NPCs) (10) or as a result of treatment with Engelbreth-Holm-Swarm gel (Matrigel®) (11). The mature SHs express genes and proteins related to hepatic differentiated functions (12). Gene expression

Takahiro Ochiya (ed.), *Liver Stem Cells: Methods and Protocols*, Methods in Molecular Biology, vol. 826,
DOI 10.1007/978-1-61779-468-1_5, © Springer Science+Business Media, LLC 2012

analysis of the SHs has revealed that CD44, which is a receptor for hyaluronic acid (HA), is specifically expressed in SHs, but not in MHs (13). Although CD44⁺ hepatocytes appear in severely injured livers, they have never been found in liver lobules of normal adult rats (14). Very recently, we have found that rat and human SHs could selectively proliferate in serum-free medium when they were cultured on HA-coated dishes (15, 16). These SHs form colonies and can be matured by treatment with Matrigel®.

In this chapter, we present the methods to separate hepatic stem/progenitor cells from the rat liver by using specific antibodies.

## 2. Materials

### 2.1. Animals

- Male F344 rats (Sankyo Lab Service, Tokyo, Japan) weighing 150–200 g (see Note 1).

### 2.2. Reagents

- Ascorbic acid-2 phosphate (Asc2P; Wako Pure Chemical Industries, Osaka, Japan, cat. No. 013-12061).
- Bovine serum albumin (30% solution; Serological Proteins, IL, cat. No. 82-046-3).
- Collagenase (Wako Pure Chemical Industries, cat. No. 034-1-533; Yakult Pharmaceutical Industry, Tokyo, Japan, cat. No. YK-101; Sigma, St Louis, MO, cat. No. G5138).
- Dexamethasone (Wako Pure Chemical Industries, cat-No.041-18861).
- Dulbecco's Modified Eagle's Medium (DMEM)-high glucose (Sigma, cat. No. D7777).
- Epidermal growth factor (EGF; BD Biosciences, Bedford, MA, cat. No. 3540001).
- Ethylenediaminetetraacetic acid (EDTA; SIGMA, cat. No. E5134).
- Ethyleneglycol bis(2-aminoethyl ether)tetraacetic acid (EGTA; Sigma, cat. No. E-0396).
- Fetal bovine serum (FBS; MP Biomedicals, Inc., Eschwege, Germany. cat. No. 2916754).
- D(+)-Galactosamine hydrochloride (GalN), 99% (Acros Organics, Geel, Belgium, cat. No. 160440010).
- Gentamicin solution (50 mg/ml; Sigma, cat. No. G1397).
- HANKS' balanced salt solution (HANKS; Sigma, cat. No. H9269).
- 10× Ca²⁺, Mg²⁺-free HANKS (Sigma, cat. No. H4641).
- HEPES (Dojindo, Kumamoto, Japan, cat. No. 342-01375).
- Insulin (Sigma, cat. No. I-5500).

- L-Proline (Sigma, cat. No. P5607).

- NaHCO$_3$ (Kanto Chemical, Tokyo, Japan, cat. No. 37116-00).

- Nembutal, 50 mg/ml (Dainippon Pharmaceutical, Tokyo, Japan, cat. No. 132141).

- Nicotinamide (Sigma, cat. No. N3376).

- Penicillin–streptomycin solution (Sigma, cat. No. P-4333).

- Phenol red-free HANKS (Invitrogen, California, USA, cat. No. 14025-092).

- Trypan blue (Chroma Technology, VT, cat. No. 1B187).

**2.3. Reagent Setup**

- 100 mM Asc2P stock solution (×100): Add Asc2P (2.90 g) to PBS (100 ml). Filter the solution with a 0.2-μm filter and store at 4°C in a 100-ml brown bottle until use (see Note 2).

- DMEM stock medium: Add the following reagents to a 1,000-ml beaker; DMEM (13.5 g), HEPES (4.76 g), L-proline (30 mg), penicillin–streptomycin (8 ml), doubly distilled (dd) H$_2$O up to 1,000 ml. During mixing with a magnetic stir bar, add NaHCO$_3$ (2.10 g) to adjust the pH to 7.4 with 1 N NaOH and then filter with a 0.2-μm filter. Store at 4°C until use.

- 10$^{-4}$ M Dexamethasone stock solution (×1,000): To make 10$^{-2}$ M stock solution, add dexamethasone (39.2 mg) to an autoclaved brown bottle and add 10 ml of ethanol. Then, dilute with autoclaved ddH$_2$O to make the final stock solution (10$^{-4}$ M). Store at 4°C until use (see Note 3).

- 500 mM EDTA stock solution: During mixing with a magnetic stir bar, add EDTA (46.5 g) to 200 ml ddH$_2$O in a graduated cylinder and then adjust to pH 8.0 with NaOH (1 N). Adjust to 250 ml with ddH$_2$O and filter with a 0.2-μm filter.

- 10 μg/ml EGF stock solution (×1,000): Add 10 ml of autoclaved ddH$_2$O to an EGF vial (following the manufacturer's instructions). Distribute into cryotubes (1 ml each) and store at –20°C until use.

- D(+)-Galactosamine stock solution (37.5 mg/ml): Add sterilized PBS (2.67 ml) to the D(+)-galactosamine hydrochloride vial (1.0 g) and dissolve by shaking, and then store at 4°C until use.

- 500 mg/ml insulin stock solution (×1,000): Add 100 mg of insulin to 100 ml of ddH$_2$O and then add 1.2 ml of 1 N HCl. Adjust to 200 ml with ddH$_2$O and filter with a 0.2-μm filter (see Note 4).

- MACS Buffer: Mix PBS (500 ml), 500 mM EDTA (2.0 ml), and 30% BSA (8.3 ml). Store at 4°C until use.

- 1 M Nicotinamide stock solution (×100): Add nicotinamide (12.21 g) to 100 ml of PBS. Filter with a 0.2-μm filter and store at 4°C until use (see Note 5).

- Perfusion solution: Add insulin stock solution (1 ml) to 200 ml of HANKS and warm at 37°C in a water bath until use. Add collagenase (100 U/ml) to prewarmed perfusion solution just before use and immediately dissolve by gentle shaking (see Note 6).

- Preperfusion solution: Prepare approximately 850 ml of ddH$_2$O in a 1,000-ml graduated cylinder. During mixing with a magnetic stir bar, add 10× Ca$^{2+}$, Mg$^{2+}$-free HANKS (100 ml), EGTA (190 mg), and insulin stock solution (1 ml), and then adjust pH to 7.5 with 1 M NaHCO$_3$. Adjust to 1,000 ml with ddH$_2$O and then filter with a 0.2-μm filter. Distribute into each bottle (150 ml) and store at 4°C until use.

- Preparation of culture medium: Add gentamicin (0.5 ml), nicotinamide stock solution (5.5 ml), Asc2P stock solution (5 ml), insulin stock solution (0.5 ml), EGF stock solution (0.5 ml), dexamethasone stock solution (0.5 ml), and FBS (50 ml) to DMEM stock medium (500 ml).

- 0.1% Trypan blue stock solution (×2): Add trypan blue (100 mg) to 100 ml of phenol red-free HANKS. Filter with a paper filter.

### 2.4. Antibodies

- Mouse anti-rat CD44 (BD Biosciences PharMingen, Franklin Lakes, NJ, Dilution 1:1,000).

- Mouse anti-rat Thy1.1 (Serotec, Raleigh, NC, Dilution 1:500).

- Rat anti-mouse IgG$_1$ microbeads (Miltenyi Biotec, Bergisch Gladbach, Germany, Dilution 1:50).

- Rat anti-mouse IgG$_{2a+b}$ microbeads (Miltenyi Biotec, Dilution 1:50).

### 2.5. Equipment

- Autoclaved 250-μm nylon filter net (Nippon Rikagaku Kikai, Tokyo, Japan).

- Butterfly needle, 18 G×3/4″, 1.27×19 mm (TOP, Tokyo, Japan, cat. No. 01201).

- Cell strainer, 70-μm filter (BD Falcon, cat. No. REF352350).

- Dishes – 35, 60, and 100 mm (Corning Glass Works, Corning, NY).

- 0.2 μm Filter (Mediakap-2; Spectrum Laboratories, CA, cat. No. MEM2M-02B-12 S).

- MACS Separation columns, 25 LS columns (Miltenyi Biotec, cat. No. 130-042-401).

- MidiMACS separation unit (Miltenyi Biotec, cat. No. 042-303).

- Paper filter no. 3 (Advantec, Tokyo, Japan).

- Peristatic pump (Tokyo Rika Instruments, Tokyo, Japan, RP-1000).

- Sterilized 10-cm Petri dish (glass or plastic).
- Vascular clamp, bulldog type (Fine Science Tools, Foster City, CA, cat. No. 18050-35).
- Water bath (Teitec Co., Tokyo, Japan, cat. No. Ex-B2015250).

## 3. Methods

### 3.1. Preparation of Liver-Injured Rat (2–4 Days Before the Experiment)

After light anesthesia using ether, 75 mg GalN/100 g body weight is intraperitoneally administered. In GalN-induced liver injury, severe hepatitis occurs and distinct jaundice is sometimes observed. In the liver lobules, hepatic stem and progenitor cells, oval cells, and SHs appear within 1 week (Fig. 1) (see Note 7).

### 3.2. Isolation of Thy1+ or CD44+ Cells

1. Settle the perfusion apparatus in the warmed water bath (38°C). Pour the preperfusion solution into the apparatus before the experiment and bubble it with 95% $O_2$/5% $CO_2$ gas at a flow rate of 0.5 l/min.

2. After light anesthesia with ether, anesthetize the rat with an intraperitoneal injection of Nembutal (5 mg per 0.1 ml per 100 g of body weight).

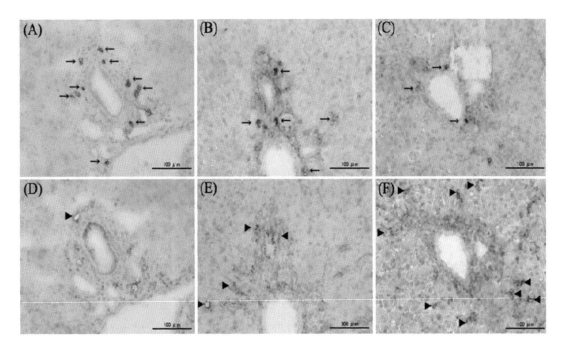

Fig. 1. Localization of Thy1+ and CD44+ cells in a GalN-injured rat liver. Serial frozen sections of GalN-treated rat livers were stained with anti-Thy1 (**A**, **B**, and **C**) and CD44 (**D**, **E**, and **F**) antibodies. The figures show 2 days (**A** and **D**), 3 days (**B** and **E**), and 4 days (**C** and **F**) after GalN administration. *Arrows* show Thy1+ cells and arrowheads CD44+ cells. Scale bars, 100 µm.

3. Cut the abdominal wall using surgical scissors and open the abdominal cavity to look for the portal vein.

4. Ligate the common bile duct and splenic vein together using a surgical thread at the portion nearest the portal vein.

5. Insert a butterfly needle filled with preperfusion solution into the portal vein 1.5–2.0 cm from the bifurcation of the portal vein, stop the tip of the needle at a position close to the bifurcation of the portal vein and clamp the needle with a surgical clip.

6. Start the perfusion at a flow rate of 30 ml/min.

7. Cut the inferior vena cava at the portion beneath the right the kidney and the thoracic cavity, and then cut the heart to flow the perfusate out of the cadaver. Washing out the blood completely from the liver and preventing an increase of intrahepatic pressure is important to the success of the preparation.

8. When the amount of the preperfusion solution becomes small, add collagenase to the perfusion solution and the pour it into the perfusion apparatus.

9. Flow the solution at a flow rate 15–20 ml/min (see Note 8).

10. Stop the flow before air bubbles move into the liver when the solution flows out from the reservoir.

11. Cut the liver from the abdominal cavity and transfer it to a sterilized Petri dish.

12. Prepare a 100-ml beaker with 70–80 ml of HANKS with insulin or medium (wash solution) and add a small amount of the wash solution to the Petri dish. From this step onward, all procedures should be done in sterilized condition.

13. Peel the hepatic capsule as carefully as possible and, to drop the digested cells, shake the liver into the beaker (see Note 9).

14. Filter the cell suspension through a 250-μm nylon filter net into a new 100-ml beaker.

15. Filter the cell suspension through a 70-μm filter, distribute the suspension into four 50-ml conical tubes using a 25-ml pipette, and then adjust each tube to an equal volume (approximately 40 ml) with the wash solution.

16. Centrifuge the tubes at $50 \times g$ for 1 min at 4°C.

17. Collect supernatants and transfer to new conical tubes. Repeat this step three times (see Note 10).

18. Centrifuge at the supernatant at $50 \times g$ for 5 min at 4°C.

19. Discard the supernatant and add 40 ml of wash solution to the tubes. Gentle pipetting is necessary to dissociate the cell pellet.

20. Centrifuge at the supernatant at $50 \times g$ for 5 min at 4°C.

21. Discard the supernatant and add 40 ml of wash solution to the tubes. Thereafter, centrifuge at $150 \times g$ for 5 min at 4°C (see Note 11).

22. Discard the supernatant, pour 20 ml of MACS buffer into each tube and gather the suspension into two 50-ml conical tubes.

23. Centrifuge the suspension at $50 \times g$ for 5 min at $4°C$.

24. Add 10 ml of MACS buffer to each tube and gather the suspension into one tube.

25. Add 0.5 ml of cell suspension to 1.5 ml of trypan blue solution and pipette gently. The cell suspension should be kept in ice.

26. Count the number of viable cells as soon as possible. For counting, use an improved Neubauer hemocytometer. Count the number of cells with trypan blue-negative nuclei. Count the number of cells that are smaller than typical MHs and larger than NPCs (around 15 µm in diameter). As many cells are dead, the overall viability may be bad. However, most SHs and oval cells are viable.

27. Centrifuge the suspension at $50 \times g$ for 5 min at $4°C$.

28. Discard the supernatant and add MACS buffer to adjust the concentration of the cells to $1 \times 10^8$ cells/ml.

29. Add 2 µg/ml anti-Thy1 or 625 ng/ml CD44 antibodies to the cell suspension and incubate on ice for 1 h.

30. Wash cells to remove unbound antibodies by adding 1–2 ml of buffer per $10^7$ cells and centrifuge the suspension at $50 \times g$ for 5 min at $4°C$.

31. Discard the supernatant and add MACS buffer to adjust the concentration of the cells to $1 \times 10^8$ cells/ml.

32. Add 200 µl of rat anti-mouse $IgG_1$ microbeads per $10^8$ cells/ml to the suspension with the anti-Thy1 antibody. In the similar way, add 200 µl of rat anti-mouse $IgG_{2a+b}$ microbeads per $10^8$ cells/ml to the suspension with the anti-CD44 antibody.

33. Mix well and incubate on ice for 30 min.

34. Wash cells to remove the unbound secondary antibody by adding 1–2 ml of MACS buffer per $10^7$ cells and centrifuge the suspension at $50 \times g$ for 5 min at $4°C$.

35. Discard the supernatant and repeat washing step 34.

36. Discard the supernatant and resuspend up to $10^8$ cells in 500 µl of MACS buffer.

37. Place an LS column in the magnetic field of the MidiMACS separation unit. Then rinse with 3 ml of MACS buffer.

38. Apply cell suspension onto the column. Collect unlabeled cells, which pass through, and wash the column with 3 ml of MACS buffer. Perform washing steps by adding MACS buffer three times, each time once the column reservoir is empty.

39. Remove the column from the separator and place it on a 15-ml conical tube.

40. Pour 5 ml of MACS buffer onto the column. Immediately flush out the fraction with the magnetically labeled cells by firmly applying the plunger supplied with the column (see Note 12).

41. Add 10 μl of cell suspension to 30 μl of trypan blue solution and pipette gently. Then, count the viable cells as soon as possible.

42. To characterize the isolated Thy1$^+$ and CD44$^+$ cells, some sorted cells are employed for analysis by RT-PCR (Fig. 2). $1 \times 10^5$ viable cells should be plated on a 12-well plate or 35-mm culture dish and cultured in the culture medium. To evaluate whether the Thy1$^+$ oval cells differentiate into SHs, immunocytochemistry for CD44 is performed to identify the SH colony (Fig. 3) (see Note 13).

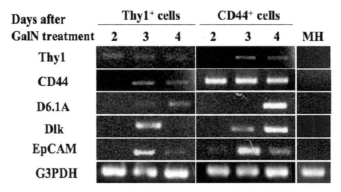

Fig. 2. Characterization of the cells isolated from a GalN-inured liver. Gene expression of markers of hepatic stem/progenitor cells was examined by RT-PCR. Total RNA was extracted from the cells sorted from GalN-treated rat livers. The numbers show the day after GalN administration.

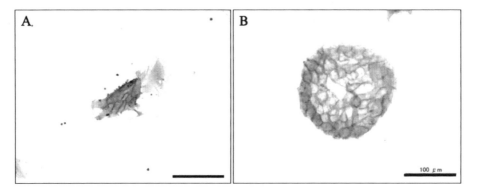

Fig. 3. Formation of SH colonies was identified by immunocyochemistry for CD44. Sorted Thy1$^+$ (**A**) and CD44$^+$ cells (**B**) from GalN-D3 rat livers were cultured for 5 days. Cell membranes were intensively stained with an anti-CD44 antibody. The size of colonies derived from CD44$^+$ cells was larger than those from Thy1$^+$ cells. Scale bars, 100 μm.

## 4. Notes

1. All animal experiments must comply with national and institutional regulations.

2. Asc2P is light sensitive and gradually inactivated even in a refrigerator.

3. $10^{-4}$ M stock solution must be used within 3 months.

4. Insulin dissolves in acidic solution.

5. Crystals are sometimes formed. As they may hurt cells, when crystals were found in the medium, fresh stock solution should be made and the culture medium should be immediately replaced. The stock must be used within 1 month.

6. Prepare HANKS with insulin before the experiment.

7. We observe that Thy1$^+$ oval cells appear in the periportal area adjacent to Glisson's capsule from 2 days after GalN administration and disappear within 1 week. On the contrary, CD44$^+$ SHs appear from 3 to 5 days after GalN administration and the appearance of CD44$^+$ cells is delayed one day compared with that of oval cells.

8. The flow rate should be decided by rat body weight.

9. A procedure of combing with a comb or tweezers may be necessary to obtain a large number of cells.

10. Many SHs and other NPCs are included in the supernatant after low-speed centrifugation. This procedure is carried out to remove the majority of MHs from the cell suspension.

11. This procedure is carried out to damage some MHs contained in the suspension.

12. To increase the purity of the magnetically labeled fraction, it can be passed over a new, freshly prepared column.

13. CD44$^+$ colonies are regarded as SH colonies. Sorted Thy1$^+$ (GalN-D2) cells cultured for 10 days rarely form CD44$^+$ colonies. On the contrary, some Thy1$^+$ (GalN-D3) cells form morphologically typical SH colonies. These results show that although most Thy1$^+$ (GalN-D2 cells) are not committed to a hepatic lineage, some Thy1$^+$ (GalN-Day3) cells may have already differentiated into CD44$^+$ SHs.

## Acknowledgments

We thank Ms. Minako Kuwano and Ms. Yumiko Tsukamoto for technical assistance. We also thank Mr. Kim Barrymore for help with the manuscript. This work was supported by Grants-in-Aid for Scientific Research (for N. Ichinohe, J. Kon, and for T. Mitaka) and the program for developing the supporting system for upgrading education and research (for T. Mitaka) from the Ministry of Education, Culture, Sports, Science, and Technology, the Science and Technology Incubation Program in Advanced Regions from the Japan Science and Technology Agency (for T. Mitaka), Research on Advanced Medical Technology from the Ministry of Health, Labor, and Welfare, and Labor Sciences Research Grants (for T. Mitaka), and the Suhara Memorial Foundation (for T. Mitaka).

## References

1. Sell S: The role of progenitor cells in repair of liver injury and in liver transplantation. Wound Rep Reg 2001, 9: 467–482

2. Forbes S, Vig P, Poulsom R, Thomas H, Alison M. Hepatic stem cells: J Pathol 2002, 197:510–518

3. Fausto N, Campbell JS: The role of hepatocytes and oval cells in liver regeneration and repopulation. Mech Dev 2003, 120:117–130

4. Roskams TA, Libbrecht L, Desmet VJ: Progenitor cells in diseased human liver. Semin Liver Dis 2003, 23:385–396

5. Knight B, Matthews VB, Olynyk JK, Yeoh GC: Jekyll and Hyde: evolving perspectives on the function and potential of the adult liver progenitor (oval) cell. BioEssays 2005, 27: 1192–1202

6. Walkup MH, Gerber DA: Hepatic stem cells: in search of. Stem Cells 2006, 24:1833–1840

7. Farber E: Similarities in the sequence of early histological changes induced in the liver of the rat by ethionine, 2-acetylaminofluorene, and 3'-methy-4-dimethylaminoazobenzene. Cancer Res 1956, 16:142–149

8. Petersen BE, Goff JP, Greenberger JS, Michalopoulos GK: Hepatic oval cells express the hematopoietic stem cell marker Thy-1 in the rat. Hepatology 1998, 27:433–45

9. Mitaka T, Kojima T, Mizuguchi T, Mochizuki Y: Growth and maturation of small hepatocytes isolated from adult liver. Biochem Biophys Res Commun 1995, 214:310–317

10. Mitaka T, Sato F, Mizuguchi T, Yokono T, Mochizuki Y: Reconstruction of hepatic organoid by rat small hepatocytes and hepatic nonparenchymal cells. Hepatology 1999, 29: 111–125

11. Sugimoto S, Mitaka T, Ikeda S, Harada K, Ikai I, Yamaoka Y, Mochizuki Y: Morphological changes induced by extracellular matrix are correlated with maturation of rat small hepatocytes. J Cell Biochem 2002, 1:16–28

12. Mitaka T, Ooe H. Drug Metabolism Reviews focusing on drug transporter interactions in the liver: characterization of hepatic-organoid cultures. Drug Metab Rev 2010, 42:472–481

13. Kon J, Ooe H, Oshima H, Kikkawa Y, Mitaka T: Expression of CD44 in rat hepatic progenitor cells. J Hepatol 2006, 45:90–98

14. Kon J, Ichinohe N, Ooe H, Chen Q, Sasaki K, Mitaka T. Thy1-positive cells have bipotential ability to differentiate into hepatocytes and biliary epithelial cells in galactosamine-induced rat liver regeneration. Am J Pathol 2009, 175:2362–2371

15. Chen Q, Kon J, Ooe H, Sasaki K, Mitaka T: Selective proliferation of rat hepatocyte progenitor cells in serum-free culture. Nat Protoc 2007, 2:1197–1205

16. Sasaki K, Kon J, Mizuguchu T, Chen Q, Ooe H, Oshima H, Hirata K, Mitaka T: Proliferation of hepatocyte progenitor cells isolated from adult human livers in serum-free medium. Cell Transplant. 2008, 17:1221–1230

# Part II

## Hepatic Differentiation from Stem Cells

# Chapter 6

# Purification of Adipose Tissue Mesenchymal Stem Cells and Differentiation Toward Hepatic-Like Cells

## Agnieszka Banas

## Abstract

There is a great interest in the development of functional hepatocytes in vitro from different types of stem cells. Multipotential mesenchymal stem cells (MSC) compose a great source for stem cell based therapy, especially, because they can be obtain from patients own tissues, sidestepping immunocompatibility and ethical issues. Among MSCs from different sources, adipose-tissue-derived mesenchymal stem cells (AT-MSCs) are very promising because of their high accessibility, proliferation ability, potentiality, and immunocompatibility.

AT-MSCs can be easily isolated from stroma vascular fraction (SVF) of adipose tissue. They represent a heterogeneous population of cells. The precise AT-MSCs's marker profile has not been defined yet; therefore, it is still not obvious how to purify these heterogeneous fraction of cells. We postulate that one of the markers defining MSC provenance is CD105 (endoglin).

Therefore, we have sorted CD105+ fraction of AT-MSCs, expanded them, and differentiated toward hepatic-like cells. In order to check their potentiality, we have firstly differentiated sorted CD105+ AT-MSCs toward mesoderm lineages, using commercialized protocols.

We have shown here, that pure CD105+ AT-MSCs fraction revealed higher homogeneity and differentiation potential toward adipogenic, osteogenic, and chondrogenic lineages and highly inducible into the hepatogenic lineage.

Generated (by using our hepatic differentiation protocol) CD105+ AT-MSCs-derived hepatic-like cells expressed hepatocyte markers, enzymes, and functions.

**Key words:** Mesenchymal stem cells, Adipose-tissue-derived mesenchymal stem cells, Adipose tissue, Liver disease, Liver regeneration, Plasticity, Differentiation

## 1. Introduction

Mesenchymal stem cells from adipose tissue (AT-MSC), so-called processed lipoaspirate (PLA) cells (1), adipose-derived stromal cells (ADSCs) (2), adipose-derived adherent stromal cells/adipose-derived adult stem cells (ADASs) (3, 4), and

Takahiro Ochiya (ed.), *Liver Stem Cells: Methods and Protocols*, Methods in Molecular Biology, vol. 826,
DOI 10.1007/978-1-61779-468-1_6, © Springer Science+Business Media, LLC 2012

adipose-tissue-derived stromal cells (ATSCs) (5) are present in a heterogeneous population of stromal vascular fraction in adipose tissue.

AT-MSCs are characterized as CD105$^+$, SH3$^+$, CD29$^+$, CD44$^+$, CD71$^+$, CD90$^+$, CD106$^+$, CD120a$^+$, CD124$^+$, CD14$^-$, CD31$^-$, CD34$^{-/+}$, and CD45$^-$, yet the surface marker profile reveals donor-to-donor variations (6–8).

One of the markers defining MSC provenance is CD105 (endoglin) (9–11). The CD105 is a component of the receptor complex of the transforming growth factor (TGF)-beta superfamily, a pleiotropic cytokine involved in cellular proliferation, differentiation, and migration. There are reports showing that CD105$^+$ – bone marrow mesenchymal stem cells (BM-MSCs) display more colony-forming unit-fibroblasts (CFU-Fs), as well as reveal a capacity to form bone in vivo (12) and a capacity to differentiate into a chondrogenic lineage (13). Additionally, the adipogenic and myogenic differentiation ratio of CD105$^+$ BM-MSCs was not influenced by the age of the donor, whereas the ratio usually decreases with patient age (14). The CD105 is also considered to be the marker of long-term repopulating hematopoietic stem cells (15).

The magnetically activated cell-sorting (MACS) of CD105$^+$ fraction was performed to obtain a multipotent and homogeneous subpopulation of cells (16). To evaluate the potentiality of CD105$^+$fraction of AT-MSCs, differentiation into adipogenic, osteogenic, and chondrogenic lineages was performed, and their differentiation ability compared with nonfractionated AT-MSCs (Fig. 1).

In order to induce CD105$^+$ AT-MSCs toward hepatic-like cells, we have used the hepatic induction strategy (16), which was based on the previously developed hepatic induction system: HIFC (Hepatic Induction Factor Cocktail), which had been established on mouse embryonic stem (ES) cells (17–19). Differentiation strategy involved the usage of growth factors: HGF, FGF1, FGF4, OsM, DEX, and collagen type I-coated dishes. After induction, CD105$^+$ AT-MSC-derived hepatic-like cells were evaluated for hepatocyte-specific morphology (Fig. 2), markers, and functions. The results indicated that CD105$^+$ AT-MSCs were highly inducible into the hepatic lineage, and derived hepatocyte-like cells expressed hepatocyte markers, proteins (Fig. 3), and functions (albumin production (Fig. 4a), ammonia detoxification (Fig. 4b) (16, 20–24).

## 2. Materials

### 2.1. Isolation of Mesenchymal Stem Cells from Human Adipose Tissue

1. Dulbecco's phosphate-buffered saline without calcium and magnesium (PBS(–)).

2. ES Cell Qualified Fetal Bovine Serum (FBS).

3. Antibiotic–antimycotic.

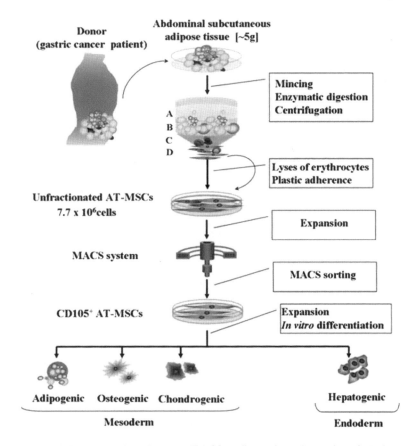

Fig. 1. Isolation and sorting of human AT-MSCs. Adipose tissue (approximately 5 g) was obtained from gastric cancer patients undergoing gastrectomy. After mincing, enzymatic digestion and centrifugation, adipose tissue separated into liquid lipid fraction (**a**), adipocyte/debris fraction (**b**), washing media fraction (**c**), and stromal vascular faction (SVF) (**d**). The SVF, the mixture of cells composed of stroma and vasculature of adipose tissue, also contains blood cells, which were washed away 24 h after plating. The remaining adherent heterogeneous fraction of cells is considered as AT-MSCs. After expansion, AT-MSCs were sorted using the MACS system, and the resulting homogeneous CD105$^+$ fraction of AT-MSCs was further analyzed and induced toward chondrogenic, adipogenic, and osteogenic lineages of mesoderm and hepatogenic lineage of endoderm.

4. Collagenase type I.

5. Scalpels, scissors, knives.

6. T-75 flask: 75 cm$^2$.

7. 160 mM NH$_4$Cl.

**2.2. Culturing and Expansion of Human Mesenchymal Stem Cells from Adipose Tissue (AT-MSCs)**

1. Dulbecco's Modified Eagle's Medium: Ham's nutrient mixture F-12 (1:1), (DMEM/F-12).

2. ES Cell Qualified Fetal Bovine Serum (FBS).

3. Antibiotic–antimycotic.

4. Culture dishes: T-75 flask: 75 cm$^2$, or ψ100 mm; 56 cm$^2$, or ψ60 mm; ψ21 cm$^2$.

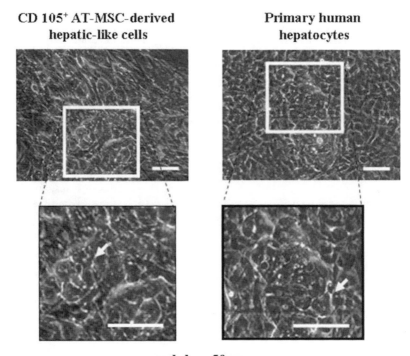

CD 105+ AT-MSC-derived hepatic-like cells          Primary human hepatocytes

scale bar: 50μm

Fig. 2. Morphological characteristics of CD105+ AT-MSC-derived hepatic-like cells (phase contrast). The morphology of CD105+ AT-MSC-derived hepatic-like cells represents many similarities with primary human hepatocytes. *Arrows* indicate bile canaliculi structures. Scale bars represent 50 μm.

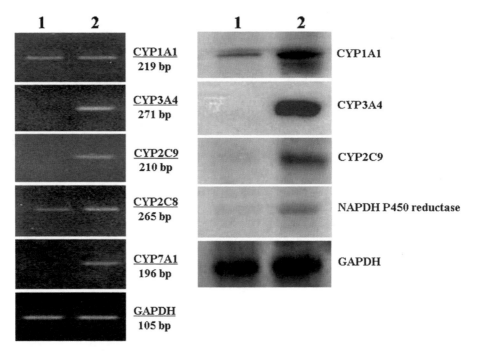

Fig. 3. RT-PCR and western blotting analyses of CD105+ AT-MSC-derived hepatic-like cells. (**a**) RT-PCR analyses of CYPs: 1A1, 3A4, 2C9, 2C8, and 7A1 in undifferentiated CD105+ AT-MSCs (*line 1*) and CD105+ AT-MSC-derived hepatic-like cells at day 40 of hepatic induction (*line 2*). (**b**) Western blot analyses of protein expression of CYPs: 1A1, 3A4, 2C9, and NADPH P-450 reductase in CD105+ AT-MSCs (*line 1*), and CD105+ AT-MSC-derived hepatic-like cells at day 50 of hepatic induction (*line 2*).

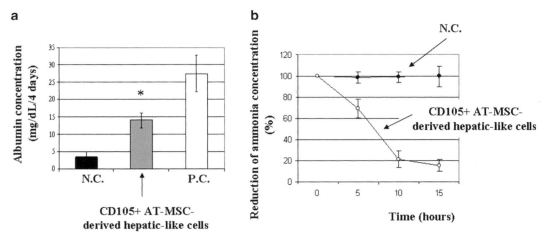

Fig. 4. Functional analyses of AT-MSC-derived hepatic-like cells (**a**) Albumin production of CD105$^+$ AT-MSC-derived hepatic-like cells (*gray filled square*) during hepatogenic induction at day 50. As a positive control (P.C) primary human hepatocytes (*open square*) were used. Negative control (N.C.) – undifferentiated CD105$^+$ AT-MSCs (*filled square*) did not reveal ability to produce albumin. (**b**)Ammonia detoxification ability of CD105$^+$ AT-MSC-derived hepatocyte-like cells (*open circle*), shown as percentage (%) of ammonia concentration. Undifferentiated CD105$^+$ AT-MSCs (N.C.) did not reveal ability to clear ammonia from the culture medium (*filled circle*). Data are reported as the mean ± SD and were analyzed by the Student's *t*-test, $n = 3$ (*$p < 0.05$).

5. Trypsin–EDTA 0.05% (1×).

6. Dulbecco's phosphate-buffered saline without calcium and magnesium (PBS(–)).

7. Tubes (15 mL).

8. Hemocytometer.

***2.3. MACS Sorting of CD105$^+$ Fraction of AT-MSCs***

1. Miltenyi Biotec Auto MACS (program Deplete).

2. MACS buffer (Running buffer): 2 mM EDTA in 0.5%BSA/PBS(–).

3. Rising buffer: 2 mM EDTA in PBS(–).

4. Anti-human CD105 magnetic microbeads (Miltenyi Biotec).

5. Trypsin–EDTA 0.05% (1×).

6. Dulbecco's phosphate-buffered saline without calcium and magnesium ((PBS(–)).

7. Tubes (15 mL).

8. Hemocytometer.

9. Dulbecco's Modified Eagle's Medium: Ham's nutrient mixture F-12 (1:1), (DMEM/F-12).

**2.4. Adipogenic, Osteogenic and Chondrogenic Differentiation of AT-MSCs/CD105+ AT-MSCs (Performed Using Commercially Available Kits)**

1. Differentiation Media BulletKit-Adipogenic (Cambrex).
2. Culture dishes: ψ60 mm; ψ21 cm².
3. Oil red O.
4. Dulbecco's phosphate-buffered saline without calcium and magnesium ((PBS(−)).
5. 10% formaldehyde.
6. Isopropanol.

*2.4.1. Adipogenic Differentiation and Oil Red O Staining*

*2.4.2. Osteogenic Differentiation and Alkaline Phosphatase Staining*

1. Differentiation Media BulletKits-Osteogenic (Cambrex).
2. Dulbecco's phosphate-buffered saline without calcium and magnesium ((PBS(−)).
3. 10% formaldehyde.
4. Culture dishes: ψ60 mm; ψ21 cm².
5. Alkaline Phosphatase Substrate Kit I (Vector).

*2.4.3. Chondrogenic Differentiation and Alcian Blue Staining*

1. Differentiation Media BulletKits – Chondrogenic (Cambrex) and TGF-$\beta$3 (final concentration 10 ng/mL) (Cambrex).
2. Polypropylene tubes.
3. Dulbecco's phosphate-buffered saline without calcium and magnesium ((PBS(−)).
4. Alcian Blue (in this case staining was performed by SRI Communication for the Health Company, Tokyo, Japan).

**2.5. Hepatic Differentiation of AT-MSCs/CD105+ AT-MSCs**

1. Hepatic Culture Medium (HCM)-modified William E medium (Cambrex), containing: transferrin (5 μg/mL), hydrocortisone-21-hemisuccinate ($10^{-6}$ mol/L), bovine serum albumin (0.5 mg/mL), ascorbic acid (2 mmol/L), epidermal growth factor (EGF) (20 ng/mL), insulin (5 μg/mL), gentamicin (50 μg/mL) (Cambrex, single quotes), and dexamethasone ($10^{-8}$ M) (Sigma).
2. Growth factors: FGF1 (300 ng/mL), FGF4 (25 ng/mL), HGF (150 ng/mL) and oncostatin M (30 ng/mL) (OsM) (PeproTech EC, London UK), and dexamethasone ($2 \times 10^{-5}$ M) (Sigma).
3. Culture dishes: Collage Type I-coated dishes: ψ60 mm; ψ21 cm², Collagen Type I 6-well plates ψ9.5 cm², Collagen Type I 24-well plate; ψ2 cm².
4. Dulbecco's phosphate-buffered saline without calcium and magnesium ((PBS(−)).
5. Collagenase type I (0.05%) – Dispase (1,000 U/mL) solution dissolved in PBS (−).

*2.5.1. Albumin Production (Fig. 4a)*

1. Albumin evaluation in medium by using the Bromocresol Green (BCG) method in this case was performed by SRI Communication for the Health Company, Tokyo, Japan.

*2.5.2. Ammonia Detoxification (Fig. 4b)*

1. Ammonia (Modified Fujii-Okuda method) (Wako).

*2.5.3. Primers and Conditions for RT-PCR*

| GenBank accession no. | Primer sequence (F: Forward; R: Reverse) | Annealing temperature (°C) | Cycles | PCR product size (bp) |
|---|---|---|---|---|
| CYP1A1 | F: TCGTCTTGGA CCTCTTTGGA R: ACGAAGGAA GAGTGTCGGAA | 58 | 43 | 219 |
| CYP2C8 | F: TGGCATTAC TGACTTCCGTG R: CCCTTTGGT AACTGCAGTAG | 58 | 43 | 265 |
| CYP2C9 | F: ACTTTCTGGA TGAAGGTGGC R: GGCACAGAGG CAAATCCATT | 58 | 43 | 210 |
| CYP3A4 | F: AGCAGAAACT GCAGGAGGAA R: TTCAGGGAGG AACTTCTCAG | 58 | 43 | 271 |
| CYP7A1 | F: AGGACGGTTC CTACAACATC R: CGATCCAAA GGGCATGTAGT | 56 | 45 | 196 |
| GAPDH | F: GAAGGTGAA GGTCGGAGT R: GAAGATGGT GATGGGATTTC | 56 | 28 | 200 |

*2.5.4. Western Blotting*

1. Rabbit anti-human CYP3A4 (Biomol).
2. Rabbit anti-human CYP2C9 (Fitzgerald).
3. Anti Rat NADPH P50 Reductase (Daiichi Pure Chemicals Co, Ltd.).
4. Mouse anti-human CYP1A1 (B-4) (Santa Cruz).
5. Goat anti-human GAPDH (V-18) (Santa Cruz Biotechnology).
6. Rabbit anti-goat IgG HRP-conjugated (Southern Biotechnology Associates. Inc.).

# 3. Methods

## 3.1. Isolation of Mesenchymal Stem Cells from Human Adipose Tissue

1. Mince adipose tissue with surgical scissors and/or knife into less than 3-mm pieces, collect them into a tube, into which add an equal volume of PBS (−), and mix vigorously at room temperature (see Note 1).

2. Let stand the mixture for a few minutes without shaking at room temperature. While setting, the mixture separates into two phases.

3. Collect the upper one (containing stem cells, adipocytes, PBS, and blood) into a new tube and wash with fresh PBS (−) three times, and discard the lower phase (containing lipid fraction).

4. Treat the upper phase with an equal volume of collagenase type I (0.075%), dissolved in PBS (−). Shake the mixture in a water bath for 30 min at 37°C.

5. After incubation, add an equal volume of DMEM containing 10% of FBS, shake the mixture well, and set for 10 min at room temperature. The solution separates into two phases.

6. Discard the upper phase (containing lipids and debris) and centrifuge the lower one (containing SVF, blood, stem cells) at $280 \times g$ for 5 min at 20°C.

7. Resuspend the cellular pellet (containing SVF, blood, stem cells) in 160 mM $NH_4Cl$ for 3 min.

8. Then, filter the mixture using a 40-μm filter into a new tube containing 5 mL of DMEM with 10% FBS.

9. Centrifuge the mixture at $280 \times g$ for 5 min at 20°C.

10. Dissolve the cell pellet, considered as stroma vascular fraction (SVF) in DMEM containing 10% FBS, and 1% antibiotic–antimycotic.

11. Count the cells and plate them at $1.0–5.0 \times 10^6$ cells per T-75 flask density, and passage them at 70–80% confluence (see Note 2).

## 3.2. Culturing and Expansion of Human Mesenchymal Stem Cells from Adipose Tissue (AT-MSCs)

1. Culture the AT-MSCs in DMEM/F12 medium, supplemented with 10% FBS and 1% antibiotic–antimycotic, in 37°C/5%$CO_2$ and a humid atmosphere.

2. When the cells reach 70–80% confluency, passage them (approximately every 3–4 days).

3. To replace the cells, wash them with PBS (−) two times and incubate for 1–2 min with 0.5 mL/21 $cm^2$ of 0.05% trypsin–EDTA at 37°C.

4. After tapping the dish, collect the cells into a new tube together with fresh DMEM/F12/10% FBS/1% antibiotic–antimycotic

medium and centrifuge at $260 \times g$ for 5 min at room temperature.

5. The ideal cell concentration for replating is $4.0–5.0 \times 10^4$ (i.e., $1 \times 10^5$ cells/60 mm dish). Count the cells using a hemocytometer.

### 3.3. MACS Sorting of CD105+ Fraction of AT-MSCs

1. Trypsinize isolated and expanded AT-MSCs, preferably at early (3–5) passage (see Note 3).

2. Suspend the cells in a MACS buffer (PBS/0.5% BSA/2 mM EDTA) and incubate them with anti-human CD105 conjugated magnetic beads for 15 min at 8°C.

3. After rinsing with the MACS buffer and centrifugation at $260 \times g$ for 10 min, separate the cells on a magnetic column previously rinsed with rising buffer.

4. After separation, one may obtain approximately $2.0–8.0 \times 10^5$ CD105+ cells.

5. Plate the MACS sorted CD105+ cells and expand them.

### 3.4. Adipogenic, Osteogenic, and Chondrogenic Differentiation of AT-MSCs/CD105+ AT-MSCs

#### 3.4.1. Adipogenic Differentiation and Oil Red O Staining

Plate the cells at concentration: $5 \times 10^5$ on 60-mm dishes. One should perform three cycles of induction/maintenance (using Cambrex differentiation medium). Each cycle consisted of 3 days of culture with Induction Medium followed by 3 days of culture in Maintenance Medium. The intensity of adipogenic induction may be noticed by microscope observation of lipid vacuoles in the induced cells, which can be stained with oil red O.

#### 3.4.2. Osteogenic Differentiation and Alkaline Phosphatase Staining

Plate the cells in 60 mm dishes ($1 \times 10^5$ cells/dish). Osteogenesis is induced by culturing the cells with Osteogenic Induction Medium for 3 weeks. Alkaline phosphatase staining indicates the osteogenic differentiation.

#### 3.4.3. Chondrogenic Differentiation and Alcian Blue Staining

Chondrogenic differentiation was performed in pellets in polypropylene culture tubes. To perform chondrogenic differentiation, $2.5 \times 10^5$ of cells were transferred to an appropriate tube, washed two times with hMSC Incomplete Chondrogenic Medium, and centrifuged at $150 \times g$ for 5 min at room temperature. The cells were resuspended in hMSC Complete Chondrogenic Medium at a concentration of $2.5 \times 10^5 / 0.5$ mL/tube and centrifuged at $150 \times g$ for 5 min at room temperature. The cell pellet was fed every 2–3 days with 0.5 mL of freshly prepared hMSC Complete Chondrogenic Medium. After 3 weeks, the pellets were fixed with formalin, embedded in paraffin, sectioned, and stained with Alcian Blue for glycosaminoglycan.

**3.5. Hepatic Differentiation of AT-MSCs/CD105⁺ AT-MSCs**

1. Plate the cells on collagen type I-coated dishes at a concentration of $3.0–4.0 \times 10^4$ cells/cm$^2$ (see Note 4).

2. When the cells reach confluency, hepatogenic induction was performed over a period of 5 weeks.

3. At first, the HCM medium was supplemented with HGF (150 ng/mL), FGF1 (300 ng/mL), and FGF4 (25 ng/mL), and the cells were cultivated with these growth factor for 2 weeks.

4. For the next 2 weeks, the cells were treated with OsM (30 ng/mL) and dexamethasone ($10^{-5}$ mol/L) and finally were cultured in HCM alone for 1–3 weeks. For some analyses, such as CYP activity assay, the cells were harvested by treatment with collagenase (0.05%)/dispase (1,000 U/mL) solution for 3–5 min and replated at a concentration of $1.0–2.0 \times 10^5$ cells/well of 24-well collagen (type I)-coated plates. The hepatogenic morphology (Fig. 2), markers (Fig. 3) and functions (Figs. 4a and 4b) were evaluated and compared with primary human hepatocytes (see Note 5).

*3.5.1. Albumin Production (Fig. 4a)*

Albumin level in the culture medium was evaluated using the Bromocresol Green (BCG) method performed by SRI Communication for the Health Company, Tokyo, Japan.

*3.5.2. Ammonia Detoxification (Fig. 4b)*

The cells were cultured in the presence of 2.5 mM NH$_4$Cl and incubated for 30 h. At 9th, 19th, and 30th hour of incubation, the medium was collected and tested for the concentration of NH$_4$Cl using Ammonia-Test.

# 4. Notes

1. Adipose tissue-derived mesenchymal stem cells (AT-MSCs) can be isolated from any subcutaneous adipose tissue. Abdominal subcutaneous adipose tissue is abundant in MSCs and can be easily obtained either during lipoaspiration or during any surgical operation. For experimental usage, ~5 g of adipose tissue is completely sufficient. In this protocol, all proportions are for starting material of ~5 g of adipose tissue.

2. From ~5 g of adipose tissue, after 7–10 days after plating, $7.7 \times 10^6$ cells may be obtained for primary culture (approximately; $1.0 \times 10^5$ to $4.6 \times 10^6$ cells/1 g of adipose tissue).

3. For MACS sorting procedure minimum $0.5–1.0 \times 10^8$ cells is required.

4. For this experiment, we used the cells at passage 5–10. As for unfractionated AT-MSCs, freshly isolated and plated AT-MSCs are more heterogeneous.

5. The hepatic induction strategy has been improved and shortened as demonstrated (25).

## Acknowledgments

We would like to thank Dr. Satoshi Suzuki (Human and Animal Bridging Research Organization), Dr. Kazunori Aoki, Ms. Nachi Namatame, Ms. Shinobu Ueda, Dr. Akihiro Kobayashi, Ms. Ayako Inoue, Fumitaka Takeshita, and Ms. Maho Kodama (National Cancer Center Research Institute) for their valuable advice and assistance. This work was supported in part by a Grant-in-Aid for the Third-Term Comprehensive 10-year Strategy for Cancer Control; Health Science Research Grants for Research on the Human Genome and Regenerative Medicine from the Ministry of Health, Labor, and Welfare of Japan; and a Grant from Japan Health Sciences Foundation.

## References

1. Zuk P.A., Zhu M., Ashjian P., De Ugarte D.A., Huang J.I., Mizuno H., Alfonso Z.C., et al. (2002) Human adipose tissue is a source of multipotent stem cells. *Mol Biol Cell* **13**, 4279–4295.

2. Strem B.M., Hicok K.C., Zhu M., Wulur I., Alfonso Z.C., Schreiber R.E., Fraser J.K., et al. (2005) Multipotential differentiation of adipose tissue-derived stem cells. *Keio J Med* **54**, 132–141.

3. Katz A.J., Tholpady A., Tholpady S.S., Shang H., Ogle R.C. (2005) Cell surface and transcriptional characterization of human adipose-derived adherent stromal (hADAS) cells. *Stem Cells* **23**, 412–423.

4. Guilak F., Lott K.E., Awad H.A., Cao Q., Hicok K.C., Fermor B., Gimble J.M. (2006) Clonal analysis of the differentiation potential of human adipocyte-derived adult stem cells. *J Cell Physiol* **206**, 229–237.

5. Lee R.H., Kim B., Choi I., Kim H., Choi H.S., Suh K., Bae Y.C., et al. (2004) Characterization and expression analysis of mesenchymal stem cells from human bone marrow and adipose tissue. *Cell Physiol Biochem* **14**, 311–324.

6. Wagner W., Wein F., Seckinger A., Frankhauser M., Wirkner U., Krause U., Blake J., et al (2005) Comparative characteristics of mesenchymal stem cells from human bone marrow, adipose tissue, and umbilical cord blood. *Exp Hematol* **33**, 1402–1416.

7. Kern S., Eichler H., Stoeve J., Kluter H., Bieback K. (2006) Comparative analysis of mesenchymal stem cells from bone marrow, umbilical cord blood or adipose tissue. *Stem Cells* **24**, 1294–1301.

8. Boquest A.C., Shahdadfar A., Fronsdal K., Sigurjonsson O., Tunheim S.H., Collas P., Brinchmann J.E. (2005) Isolation and transcription profiling of purified uncultured human stromal stem cells: alteration of gene expression after in vitro cell culture. *Mol Biol Cell* **16**, 1131–1141.

9. Pittenger M.F, Mackay A.M., Beck S.C., Jaiswal R.K., Douglas R., Mosca J.D., Moorman M.A., et al. (1999) Multilineage potential of adult human mesenchymal stem cells. *Science* **284**, 143–147.

10. Haynesworth S.E., Baber M.A., Caplan A.I. (1992) Cell surface antigens on human marrow-derived mesenchymal stem cells are detected by monoclonal antibodies. *Bone* **13**, 69–80.

11. Liu P.G., Zhou D.B., Shen T. (2005) Identification of human bone marrow mesenchymal stem cells: preparation and utilization of two monoclonal antibodies against SH2,

SH3. (in Chinese) *Zhongguo Shi Yan Xue Ye Xue Za Zhi* **13**, 656–659.

12. Aslan H., Zilberman Y., Kendel A., Liebergall M., Oskouian R.J., Gazit D., Gazit Z. (2006) Osteogenic differentiation of noncultured immunoisolated bone marrow-derived CD105+ cells. *Stem Cells* **24**, 1728–1737.

13. Majumdar M.K., Banks V., Peluso D.P., Morris E.A. (2000) Isolation, characterization, and chondrogenic potential of human bone marrow-derived multipotential stromal cells. *J Cell Physiol* **185**, 98–106.

14. Roura S., Farre J., Soler-Botija C., Llach A, Hove-Madsen L., Cairo J.J., Godia F., et al. (2006) Effect of aging on the pluripotent capacity of human CD105+ mesenchymal stem cells. *Eur J Heart Fail* **8**, 555–563.

15. Chen C.Z., Li M., de Graaf D., Monti S., et al. (2002) Identification of endoglin as a functional marker that defines long-term repopulating hematopoietic stem cells. *Proc Natl Acad Sci USA* **99**, 15468–15473.

16. Banas A., Teratani T., Yamamoto Y., Tokuhara M., Takeshita F., Quinn G., Okochi H., Ochiya T. (2007) Adipose tissue-derived mesenchymal stem cells as a source of human hepatocytes. *Hepatology* **46**, 219–228.

17. Teratani T., Yamamoto H., Aoyagi K., Sasaki H., Asari A., Quinn G., Sasaki H., Terada M., Ochiya T. (2005) Direct hepatic fate specification from mouse embryonic stem cells. *Hepatology* **41**, 836–846.

18. Yamamoto Y., Teratani T., Yamamoto H., Quinn G., Murata S., Ikeda R., Kinoshita K., Matsubara K., Kato T., Ochiya T. (2005) Recapitulation of in vivo gene expression during hepatic differentiation from embryonic stem cells. *Hepatology* **42**, 558–567.

19. Yamamoto H., Quinn G., Asari A., Yamanokuchi H., Teratani T., Terada M., Ochiya T. (2003) Differentiation of embryonic stem cells into hepatocytes: biological functions and therapeutic application. *Hepatology* **37**, 983–993.

20. Banas A., Yamamoto Y., Teratani T., Ochiya T. (2007) Stem cell plasticity: Learning from hepatogenic differentiation strategies. *Developmental Dynamics* **236**, 3228–3241.

21. Banas A., Teratani T., Yamamoto Y., Tokuhara M., Takeshita F., Osaki M., Kawamata M., Kato T., Okochi H., Ochiya T. (2008) IFATS Collection: In vivo therapeutic potential of human adipose tissue mesenchymal stem cells after transplantation into mice with liver injury. *Stem Cells* **26**, 2705–2712.

22. Yamamoto Y., Banas A., Murata S., Teratani T., Lim C.R., Hatada I., Matsubara K., Kato T., Ochiya T. (2008) A comparative analysis of transcriptome and signal pathway in hepatic differentiation of human adipose mesenchymal stem cells. *FEBS J* **275**, 1260–1273.

23. Ishikawa T., Banas A., Hagiwara K., Iwaguro H., Ochiya T. (2010) Stem Cells for hepatic regeneration: the role of adipose tissue derived mesenchymal stem cells. *Curr Stem Cell Res Ther* **5**, 182–189.

24. Ochiya T., Yamamoto Y., Banas A. (2010) Commitment of stem cells into functional hepatocytes. *Differentiation* **79**, 65–73.

25. Banas A., Teratani T., Yamamoto Y., Tokuhara M., Takeshita F., Osaki M., Kato T., Okochi H., Ochiya T. (2009) A rapid hepatic fate specification of adipose-derived stem cells (ASCs) and their therapeutic potential for liver failure. *J Gastr Hepatol* **24**, 70–77.

# Chapter 7

# Development of Immortalized Hepatocyte-Like Cells from hMSCs

## Adisak Wongkajornsilp, Khanit Sa-ngiamsuntorn, and Suradej Hongeng

## Abstract

Clones of hepatocyte-like cells were reproducibly generated from human mesenchymal stem cells immortalized with a combined transduction of both Bmi-1 and TERT genes. These hepatocyte-like cells contained selective markers and several functional properties of hepatocytes, yet still carried proliferative potential. These cells had cuboidal morphology and arranged themselves as cord-like structure in culture. The cloned cells deposited glycogen and actively synthesized albumin. The basal expressions of CYP450 isozymes was observed, albeit only 10–20% that of primary hepatocytes. These expressions were promptly increased upon the addition of rifampicin, a known enzyme inducer. These hepatocyte-like cells may serve as a close alternative to the use of primary hepatocytes for in vitro studies.

**Key words:** Hepatocyte-like cell, hMSC, Cell immortalization, Hepatocyte differentiation, CYP450, Drug metabolism, Toxicology

## 1. Introduction

The procurement of human hepatocyte cell lines would benefit in diverse applications, namely, allotransplant, xenobiotic biotransformation, and assessment of CYP450 activation profiles. The generation of primary hepatocyte is complicated with both ethical and technical hitches. The activity of drug metabolizing enzymes and many transporter functions were rapidly lost after being cultured (1, 2). The primary human hepatocytes maintained their functions for 3 days and barely survived up to 7 days except under special condition (3–5). The primary human hepatocyte remains a gold standard for in vitro study of drug metabolism and toxicology (4). To date, there has been only a single continuous non-cancerous

Takahiro Ochiya (ed.), *Liver Stem Cells: Methods and Protocols*, Methods in Molecular Biology, vol. 826,
DOI 10.1007/978-1-61779-468-1_7, © Springer Science+Business Media, LLC 2012

human hepatocyte cell line (Fa2N-4) with a maximal induction of CYP450 transcripts was only ten times its low basal level (6, 7). The expectation of hepatocyte cell line with functional integrity is, therefore, currently not realistic and alternative cells carrying hepatocyte-emulative functions should be acceptably substituted. One of the closest examples of such cells is the hepatocyte-like cells derived from human mesenchymal stem cells (hMSCs) (8).

hMSCs could give rise to diversely specific cell types such as chondrocytes, osteocytes, adipocytes, and hepatocytes (9–12). The potential of MSCs derived from bone marrow, adipose tissue, or umbilical cord blood to differentiate into hepatocytes has long been shown in humans (13–15), using specialized growth conditions in vitro. The multipotent stem cell derived hepatocyte-like cells could be applied for the study of hepatic biotransformation of xenobiotics and hepatotoxicity (16, 17). Since hMSCs have the ability of self-renewal, the use of hMSC-derived hepatocytes would serve as an unlimited substitution for functioning human hepatocytes.

The validity for using immortalized cell line for in vitro metabolic study relies on the maintenance of hepatocyte phenotypes as represented by a panel of specific markers. These hepatocyte-like cells contain all known drug-metabolizing enzymes, including CYP450 isozymes. The precursor hMSC had been immortalized through the transduction with two entropic lentiviral plasmids separately encoding human telomerase reverse transcriptase gene (hTERT) and Bmi-1 (18). The resulting differentiated immortalized cells contained not only hepatocyte phenotypes but also proliferative activity.

## 2. Materials

### 2.1. Isolation of Mesenchymal Stem Cell

1. IsoPrep® (Robbins Scientific, Canada).
2. Improved Neubauer hemocytometer.

### 2.2. Culture of hMSCs

1. Minimum Essential Medium-α (α-MEM, Gibco/BRL, Cat. No. 12000-063) supplemented with 10% fetal bovine serum (FBS, Biochrom AG Berlin, Germany).
2. Iscove's Modified Dulbecco's Medium (IMDM) (Gibco/BRL, Cat. No. 12200-036).
3. Porcine trypsin, 0.25% w/v, 1 mM EDTA in PBS.
4. Phosphate-buffered saline (PBS) without calcium and magnesium.
5. Plastic wares (polypropylene centrifuge tubes 15 and 50 mL, plastic tissue culture Petri dishes 10-cm diameter, cell culture flash T-25 and T-75 (Corning Incorporated, USA), 6-cm

diameter collagen IV-coated culture dishes (Iwaki Glass Co., Tokyo, Japan).

6. Penicillin G sodium (Sigma, MO, Cat. No. P7794).

7. Streptomycin (Sigma, MO, Cat. No. S6501).

8. Trypan Blue solution, 0.85% in saline (Trypan blue stain (Sigma, MO).

9. Trypsin 250 (Difco Laboratories, USA).

**2.3. The Immortalization of hMSCs**

1. Lentivirus plasmid vector 12245: pLOX-TERT-iresTK, plasmid 12240: pLOX-CWBmil, plasmid 12260: psPAX2 and plasmid 12259: pMD2.G (Addgene, Inc., USA).

2. Luria Broth (LB, 500 mL): 5 g tryptone, 2.5 g Yeast extract, 5 g NaCl and 500 mL $H_2O$. Autoclave using liquid cycle and store at 4°C.

3. QIAGEN Plasmid Midi Kit (Qiagen, Germany, Cat. No. 12143).

4. Sterile syringe filter (0.45 µM Sartorius, Germany).

5. Chloroquine diphosphate salt (Sigma, MO, Cat. No. C6628).

6. Hexadimethrine bromide or polybrene (Sigma, MO, No. H9268).

7. 293T human embryonic kidney cells (ATCC, Cat. No. CRL-11268).

8. 2× HEPES-buffered saline (HBS) solution (50 mM HEPES, 1.5 mM $Na_2HPO_4$, 280 mM NaCl, 10 mM KCl, 12 mM sucrose) filter-sterilize or autoclave. Solution can be stored at –20°C for at least 1 year. HEPES (Sigma, MO, Cat. No. H4034).

9. 2 M $CaCl_2$ stock solution: Dissolve 14.7 g of $CaCl_2$ and adjust to 100 ml with $H_2O$ and filter-sterilize and store at –20°C, stable for at least 1 year. $CaCl_2$ (Sigma, MO, Cat. No. C-2536).

**2.4. The Hepatocyte Differentiation**

1. Human recombinant epidermal growth factor (EGF, Chemicon Millipore, CA, Cat. No. GF144).

2. Human recombinant basic fibroblast growth factor (bFGF, Chemicon Millipore, CA, Cat. No. GF003-AF).

3. Human recombinant hepatocyte growth factor (HGF, Chemicon Millipore, CA, Cat. No. GF116).

4. Human recombinant oncostatin M (OSM, Chemicon Millipore, CA, Cat. No. GF016).

5. Nicotinamide (Sigma, MO, Cat. No. N0636).

6. Dexamethasone – water soluble (Sigma, MO, Cat. No. D2915).

7. Insulin-Transferrin-Selenium-A (ITS) Supplement (100×) (Invitrogen, Cat. No. 51300-044).

| | |
|---|---|
| ***2.5. RNA Extraction and Quantitative Real-Time PCR*** | 1. RNA extraction RNeasy Mini kit (Qiagen, Hiden, Germany, Cat. No. 74104). |
| | 2. ImProm-II Reverse Transcription System (Madison, WI, Cat. No. A3800). |
| | 3. FastStart Universal SYBR Green Master (ROX) (Roche Diagnostics, Germany, Cat. No. 04913949001). |
| | 4. Primer sets for Bmi-1, hTERT, and hepatocyte-specific genes (see Note 1). |

***2.6. The Functional Analysis of Differentiated Cells***

QuantiChrom™ Urea Assay Kit (DIUR-500, Bioxys, Belgium).

*2.6.1. Urea Production Assay*

*2.6.2. Glycogen Storage (PAS Assay)*

Periodic Acid-Schiff (PAS) Kit (395B, Sigma, MO).

| | |
|---|---|
| *2.6.3. Albumin Accumulation* | 1. Human Serum Albumin antibody (ab2406, MA). |
| | 2. FACS Perm (BD Bioscience, CA). |
| | 3. Triton X-100 (Sigma, MO). |
| | 4. Albumin from bovine serum (Sigma, MO). |
| *2.6.4. Immunofluorescence* | 1. Goat anti-mouse IgG conjugated to FITC (Santa Cruz Biotechnology, CA). |
| | 2. Cytochrome P450 3A4 antibody (Abcam, MA). |
| ***2.7. The Induction of CYP450 Activities*** | 1. Rifampicin (Sigma, MO, Cat. No. R3501). |
| | 2. Omeprazole (Sigma, MO, Cat. No. O104). |
| | 3. Phenobarbital (Sigma, MO, Cat. No. P5178). |
| | 4. Ethanol (Sigma, MO). |
| | 5. DMSO (Sigma, MO). |
| | 6. P450-glo Luminescent Cytochrome P450 Assay CYP1A1, 1A2, 2C9, and 3A4 (Madison, WI, Cat. No. V8751, V8771, V8791, V9001). |

# 3. Methods

| | |
|---|---|
| ***3.1. The Isolation of hMSCs from Bone Marrow Aspirate (Fig. 1)*** | 1. The bone marrow aspirate (5 mL) was transferred to 50-mL centrifuge tube. The bone marrow aspirate was diluted with PBS at a ratio of 1:3 to reconstitute the volume up to 20 mL. |

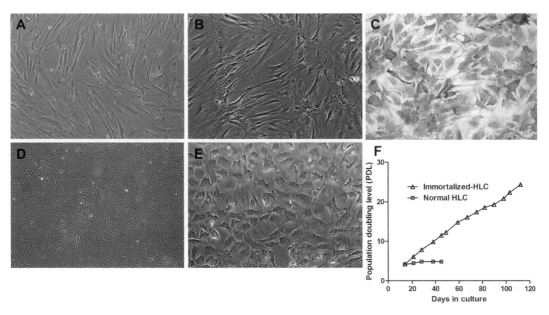

Fig. 1. The cellular morphology of hMSC and hepatocyte-like cell. The mononuclear cells from bone marrow aspirated were isolated using Ficoll gradient centrifugation (**a**). After reaching the 20th passage, hMSCs were immortalized (BMI/hTERT-hMSC, **b**). The glycogen deposit in hepatocyte-like cells in the fourth passage was demonstrated using PAS assay (**c**). The hepatocyte-like cells had cuboidal shape with cord-like arrangement right after the induction (**d**) and after being cultured in DMEM/F12 plus 10% FBS for another ten passages (**e**). The immortalized hepatocyte-like cell still contained proliferative property (**f**) as demonstrated using population doubling level (PDL).

2. To each of the two 15-mL polypropylene centrifuge tubes, 3 mL of Ficoll–Hypaque reagent (IsoPrep®) was added followed by the gently overlay of the 10 mL of bone marrow dilution over the Ficoll. The solvent phase junction between the bone marrow layer and the Ficoll layer should not be disrupted.

3. The cell suspension was centrifuged at $1,000 \times g$ for 30 min at 20°C with the break off.

4. After centrifugation, the mononuclear cell layer would appear as a white ring just above the interface between the diluted bone marrow layer and the Ficoll layer at the bottom. The ring could be collected using sterile Pasteur pipette and transferred to a new 50-mL centrifuge tube.

5. The white pellet from step 4 would be washed thrice with threefold volume of PBS and centrifuged at $1,500 \times g$ for 10 min at room temperature.

6. The viability of the cells could be assessed with trypan blue exclusion assay. The cell suspension (10 µL) was mixed with 10 µL trypan blue and laid over a hemocytometer. The cell viability should be above 80%.

7. The cell pellet was resuspended in 20 mL growth medium (α-MEM supplemented with 10% FBS, penicillin and streptomycin).

8. The cell suspension was transferred to T-75 cell culture flasks in a humidified incubator with 5% $CO_2$. The medium will be replaced first at day 4–5 followed by every 3–4 days thereafter.

9. After 7 days, the adherent cells could be assessed using an inverted microscope. Fibroblast-like colony should be presented clearly on the culture surface. The possible contamination of the culture cell with some hematopoietic cells could be eliminated through repeated passaging.

### 3.2. The Maintenance of hMSCs

From 3.1, hMSCs would reach 70–80% confluence by day 10–14. Upon reaching confluence, the hMSCs could be trypsinized and seeded as monolayer as followed:

1. The conditioned medium in T-75 culture flask will be aspirated from the adherent cells and replaced with 4 mL of trypsin/EDTA.

2. The T-75 cell culture flask would be brought into a 37°C incubator for 2–3 min, and inspected for the monolayer using an inverted microscope with 10× objective lens. The adherent cells should be detached from the culture surface. Additional incubation at 37°C for 5 min might be necessary if adherent cells did not detach well.

3. To inactivate the trypsin activity, the cell suspension was reconstituted with 4 mL of culture medium with 10% FBS, transferred to 50 mL conical centrifuge tube, and centrifuged for 10 min at $1,000 \times g$.

4. The supernatant was removed from the cell pellet. The pellet was resuspended and washed with 1–2 mL of pre-warmed culture medium.

5. A 10 μL aliquot of cell suspension was mixed with 10 μL of trypan blue and count with a hemocytometer. Cell viability should be at least 80%.

6. Cell density per T-75 flask should stay between $2 \times 10^6$ and $5 \times 10^6$ cells and incubated in 5% $CO_2$ at 37°C.

7. The culture medium should be replaced twice a week and subculture once a week.

### 3.3. Preparation of Lentivirus Vectors for Immortalization

1. Approximately 24 h before transfection, HEK293T cells ($4 \times 10^6$ cells) in 10 mL DMEM, 10% FBS, penicillin and streptomycin were seeded over a 10-cm culture dish. The dish should be gently shaken side to side to evenly distribute the cells. After adding cells, gently mix the dish up–down and left–right. The adherent cells should reach 60–70% confluence at the time of transfection.

2. Lentiviruses plasmid DNA compose of psPAX2 (Addgene plasmid 12260) packaging vector and pMD2.G (Addgene plasmid 12259) vesicular stomatitis virus G envelope, and the plasmid

encoding either hTERT (pLOX-TERT-iresTK, Addgene plasmid 12245) or Bmi-1 (pLOX-CWBmi1, Addgene plasmid 12240) were obtained by Addgene.

3. The lentivirus plasmid DNA was amplified using plasmid miniprep or midiprep from an overnight transformed *E. coli* culture grown in 10 mL LB medium. A 10 mL overnight LB culture should yield 5–10 µg DNA.

4. On the transfection date, culture medium would be gently removed from the 10-cm culture dish and replaced with 10 mL DMEM, 10% FBS, and 25 µM chloroquine.

5. In a sterile 15-mL conical tube, 10 µg pLOX-TERT-iresTK or 10 µg pLOX-CWBmi1 would be mixed thoroughly with 6.5 µg psPAX2 packaging plasmid, 3.5 µg pMD2.G vesicular stomatitis virus G envelope plasmid, 290 µL 0.1× TE buffer, 160 µL sterile $H_2O$, and 50 µL 2M $CaCl_2$. 2× HBS (500 µL) would be added drop by drop while gently mixing. The mixture was left for 5 min to allow fine precipitation.

6. Each solution condition (1-mL) from step 5 would be layered onto the 10-cm culture dish drop by drop using a micropipette to cover all culture area.

7. The culture dishes were incubated in 5% $CO_2$ at 37°C overnight (16–18 h).

8. The cells will be examined under a microscope. Cells should appear healthy and be around 80–90% confluent. A fine precipitate should be visible in culture medium. The incubation was stopped by replacing the medium with 10 mL fresh DMEM, 10% FBS. The cell culture dishes were further incubated with 5% $CO_2$ at 37°C for 48 h. Viral particles could be harvested at 48–50 h after complete incubation period.

9. After 48 h, the supernatant could be collected. The supernatants were pooled into a 50-mL conical tube and filtered thought a 0.45-µm sterile syringe filter to remove cell debris. Viral stock could be concentrated by ultracentrifugation and kept frozen at −70°C until future use.

***3.4. Immortalization of hMSC with Bmi-1 and hTERT Lentiviral Transduction***

1. hMSCs between the third and fifth passages were seeded at a density of $2 \times 10^6$ cells/mL α-MEM, 10% FBS, antibiotic onto 6-well plate.

2. Before transduction, cells should reach 60–70% confluent. The Bmi-1 and hTERT lentiviral stock (1:1, 1:2, 1:4) should be diluted with culture medium to determine the suitable MOI (multiplicity of infection, see Note 1) for MSCs. Our determined optimal ratio was 1:2. Both lentiviruses were mixed together for 2-gene transduction.

3. The lentiviral supernatant was mixed with α-MEM, 10% FBS to reconstitute as the final transfection medium. The final

transfection medium was dispensed as 1 mL/well with 6 µg/mL polybrene.

4. The incubation proceeded overnight (16–18 h) in 5% $CO_2$ at 37°C and was stopped by replacing the transfection medium with fresh culture medium. The infected MSCs were maintained for another 3 days to allow the expression of the transduced genes.

**3.5. Cloning of Immortalized Human Mesenchymal Stem Cells**

1. After transduction for 3–4 days, MSC should reach 80–90% confluent. The cells could be trypsinized and checked whether the viability were higher than 80%.

2. The cells were resuspended initially as $1 \times 10^4$ cell/mL in α-MEM medium, 10% FBS.

3. Five sterile conical centrifuge tubes were brought for sequential dilution. Each tube was filled with 3 mL fresh growth medium. Cell suspension (2 mL) from step 2 were transferred to tube number 1 and mixed using pipette.

4. A 2-mL aliquot of diluted cell suspension in tube number 1 was transferred to the tube number 2. Cell suspension was diluted and transferred to the next tube in the same manner until tube number 5. The diluted cell suspension in tube number 4 or 5 was suitable for single cell cloning.

5. Cell suspension (0.5 mL) from step 4 was transferred to each well of a sterile 24-well plate to achieve a single cell/well.

6. The 24-well plate was incubated in 5% $CO_2$ at 37°C for 1 week. The medium was replaced with fresh growth medium every 3–4 days. The culture was continued until the adherent cells derived from a single cell reached 70–80 confluent.

7. The expanded cloned cell from step 6 were trypsinized and transferred to a T-25 tissue culture flask. At least 6–8 clones were picked and screened for the highest expression of both Bmi-1 and hTERT genes.

**3.6. Quantitative Real-Time PCR Analysis for Cell-Specific Markers**

1. The adherent cells were trypsinized and washed twice with PBS by centrifugation at $1,000 \times g$ for 5 min at 4°C. The supernatant was removed while the cell pellet could be immediately used for RNA extraction or stored for a long term at –70°C. The RNA extraction was performed using RNeasy Minikit (Qiagen) following the manufacturer instruction.

2. The commercial ImProm-II™ Reverse Transcription system was used to synthesize single-stranded cDNA according to the manufacturer instruction. Briefly, 4 µL RNA template was mixed with 1 µL oligo (dT) in the first microcentrifuge tube. The tube was heated to 70°C for 5 min and then placed on ice for 5 min. A master mix containing 4.8 µL of 25mM $MgCl_2$, 4 µL of 5× reaction buffer, 3.7 µL RNase free water,

1 μL reverse transcriptase, 1 μL of 10 mM dNTPs, and 0.5 μL DNase inhibitor was added. The temperature was adjusted to 25°C for 5 min, 42°C for 1 h, 70°C for 15 min, and 4°C for 5 min.

3. The cDNA concentration was determined using the NanoDrop® spectrophotometer and diluted with double-distilled water to 10–100 μg/mL for immediate use or long-term storage at −70°C.

4. For real-time PCR, each reaction would contain 10 μL FastStart SYBR Green Master or equivalent, 7.5 μL double-distilled water, specific primer pairs (see Note 2), and 0.1 μg of cDNA from step 3. The temperature cycle in the real-time PCR (StepOnePlus®) consisted of 95°C for 10 min, followed by 40 cycles of 95°C for 15 s, 60°C for 40 s, and 72°C for 40 s.

5. For data analysis, the cycle threshold (Ct) numbers were computed for each sample using the Sequence Detection Software Version 2.01 (Applied Biosystems). To obtain accurate comparison, all hepatocyte-specific genes were normalized with the endogenous housekeeping gene, glyceraldehyde-3-phosphate dehydrogenase (GAPDH).

### 3.7. The Induction of Hepatogenesis

1. The unmanipulated hMSCs between the third and fifth passages or BMI/hTERT-transduced MSCs with a density of $1 \times 10^4$ cells/cm² were seeded for 2 days.

2. For initiation step, the hMSCs were cultured in Iscove's Modified Dulbecco's Medium (IMDM), 20 ng/mL epidermal growth factor (EGF), and 10 ng/mL basic fibroblast growth factor (bFGF) for 2 days.

3. For differentiation step, hMSCs were maintained in IMDM, 20 ng/mL HGF, 10 ng/mL bFGF, and 0.61 g/L nicotinamide for 7 days.

4. For maturation step, MSCs were maintained in IMDM, 20 ng/mL oncostatin M, 1 μM dexamethasone, and 1% (v/v) ITS for 14 days, with routine medium change every 3 days.

5. For CYP450 induction, a cocktail of prototypic CYPs inducers (i.e., 40 μM rifampicin, 50 μM dexamethasone, 1 mM omeprazole, 50 μM phenobarbital, and 0.1% (v/v) DMSO with 2% FBS) was added to the cells and incubated for 3 days with daily medium change.

### 3.8. Hepatocyte Functional Analysis (Fig. 2)

#### 3.8.1. Urea Production Assay

1. The cultured cells (hMSCs, hepatocyte-like cells or HepG2) in IMDM were incubated with 5 mM NH$_4$Cl for 48 h.

2. Either the classical diacetyl monoxime test or the commercial QuantiChrom Urea Assay Kit (DIUR-500) could be employed. The conditioned medium (5 μL) from step 1 was collected and transferred in duplicate onto each well of a clear bottom

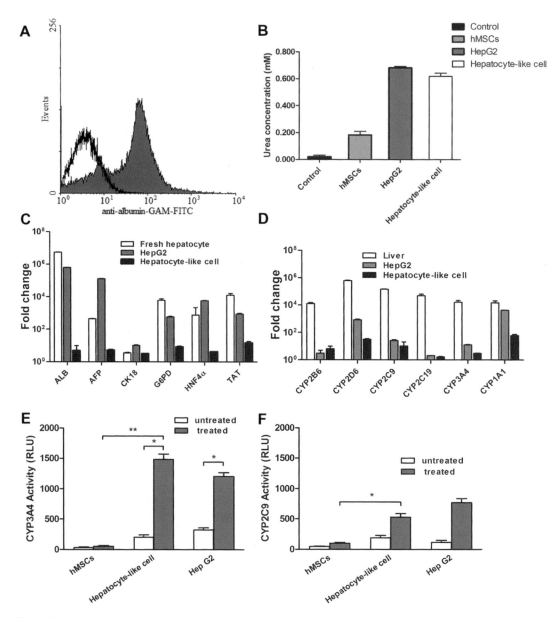

Fig. 2. The expression of hepatocyte-selective markers and functions. The hepatocyte-like cells were investigated in comparison with HepG2 for albumin synthesis using FACS (**a**); and urea production using urea assay (**b**). The hepatocyte-like cells were studied for the expression of hepatocyte-selective genes (**c**) and CYP450 (**d**) at basal stage. After the induction with 40 μM rifampicin for 72 h, the functional activity of CYP3A4 (**e**) and CYP2A9 (**f**) was assayed.

96-well plate. The blank water or urea standard (50 mg/dL) was transferred in the same fashion. The 200 μL working reagent was added to each well. The solution was mixed by tapping the plate lightly. The incubation was carried out for 10–20 min at room temperature.

3. The plate was read for absorbance at 470–550 nm in a spectrophotometer. The peak absorbance is at 520 nm.

4. Urea concentration of the sample could be calculated as:

$$\text{Urea concentration} = \frac{\text{OD}_{sample} - \text{OD}_{blank}}{\text{OD}_{standard} - \text{OD}_{blank}} \times n \times 50 (mg / dL)$$

$\text{OD}_{sample}$, $\text{OD}_{blank}$, and $\text{OD}_{standard}$ are OD520 nm of sample, water, and standard, respectively. $n$ is the dilution factor. Urea at 1 mg/dL is equal to 167 μM, 0.001%, or 10 ppm.

*3.8.2. Glycogen Synthesis (Periodic Acid-Schiff, PAS) Assay*

1. The trypsinized cultured cells (hMSCs, hepatocyte-like cells or HepG2) were transferred to collagen type I-coated coverslip. The cells were allowed to grow until reaching 80–90% confluent.

2. The coverslip was fixed with 4% paraformaldehyde for 30 min, permeabilized with 0.1% Triton X-100 for 10 min, incubated with or without diastase for 1 h at 37°C, oxidized in 1% periodic acid for 5 min, rinsed thrice with dH$_2$O, treated with PAS reagent for 15 min, and rinsed with water for 5–10 min.

3. The attached cells were counterstained with Mayer's hematoxylin for 1 min, rinsed with water, and assessed under light microscope. The resulting density of oxidized glycogen could be visualized as a color gradient starting from pink to strong red.

*3.8.3. Analysis of Cellular Markers Using Flow Cytometry*

1. The cultured cells were trypsinized and washed with PBS through centrifugation at $1,000 \times g$ for 5 min at 4°C.

2. The washed cells were resuspended with 0.5 mL FACS buffer depending on cell density.

3. The cell suspension was transferred to BD FACS tube. The primary antibody (0.5–1 μL anti-albumin, anti-CD105, or anti-CD90, etc.) was added and mixed. The stained cells were incubated at 4°C for 30–40 min, washed, and centrifuged 2–3 times.

4. If any primary antibody does not conjugated with fluorochrome, secondary antibody such as GAM-FITC, GAR-PE is needed for FACS analysis.

5. After the staining, centrifugation, and washing thrice, the cells were suspended in 500 μL FACS buffer in BD tube.

6. The suspending cells were ready for analysis using flow cytometry. Nonsingle cells and debris could be omitted based on FSC and dead cell based on SSC. At least 10,000 cells were analyzed per sample.

7. The FACS data could be analyzed using FlowJo version 7.63 or WinMDI version 2.9.

**3.8.4. Immunofluorescence Microscopy**

1. The trypsinized hepatocyte-like cells were transferred to collagen type I-coated coverslip. Cells were allowed to grow until reaching 80–90% confluent.

2. The adherent cells were washed briefly with PBS, Fix with 4% paraformaldehyde at room temperature for 30 min, followed by 100% ethanol for another 10 min.

3. The cells were washed thrice with PBS, blocked with 5% normal serum from the same species as the secondary antibody in 1% BSA/0.2% Triton X-100/PBS for 1 h at room temperature.

4. The cells were incubated with the primary antibody (anti-CYP3A4, anti-CYP2C9, or anti-CYP1A1, etc., Abcam, Cambridge, MA) for 1 h at 37°C in moist chamber.

5. After washing with PBS thrice, the cells were mounted with antifade mounting medium in coverslip and examined under a fluorescent microscope and photographed.

**3.9. The Analysis of CYP450 Activity**

1. The hepatocyte-like cell or HepG2 were incubated in growth medium supplement with prototypic inducers such as 40 μM rifampicin, 50 μM dexamethasone, 1 mM omeprazole, 50 μM phenobarbital, 50 μM artesunate for 72 h with daily medium change prior to the assay.

2. After 3-days incubation period, the cells were incubated with IMDM supplemented with 100 μM Luciferin-CEE (CYP1A1), Luciferin-H (CYP1A2), or Luciferin-ME (CYP2C9) for 3–4 h or 3 μM Luciferin-IPA (CYP3A4) for 30–60 min. A 50 μL aliquot of the incubation medium was transferred to 96-well opaque white luminometer plate. Luciferin detection reagent was added into each well.

3. The plate was incubated at room temperature for 20 min in dark chamber.

4. The luminescence was determined using a luminometer or an attached CCD camera for the measurement of luminescence unit.

5. The relative luminescence unit (RLU) could be calculated as follows:

$$RLU = \frac{LU_{treated} - LU_{blank}}{LU_{untreated} - LU_{blank}}$$

# 4. Notes

1. The ratio of lentiviral supernatant to culture medium was 1:2 or 1:4 for the immortalization of hMSCs. The quantitation of living viral stocks was required to determine the exact multiplicity of infection (MOI). Freshly harvested viral stocks can be

quantitated immediately, or frozen in aliquots at –80°C for later measurement. Each freeze–thaw cycle could reduce the functional titer of the viral stock up to two- to fourfolds. The MOI is heavily relied on the cell types and measuring methods. The MOI could be determined with quantitative PCR or flow cytometry (19).

2. Primer for real-time PCR analysis

| Gene | Forward primer | Reverse primer | Amplicon (bp) |
|---|---|---|---|
| ALB | TGAGAAAACG CCAGTAAGTGAC | TGCGAAATCATC CATAACAGC | 265 |
| AFP | GCTTGGTGGT GGATGAAACA | TCCTCTGTTATTT GTGGCTTTTG | 157 |
| CK18 | GAGATCGAGG CTCTCAAGGA | CAAGCTGGCCT TCAGATTTC | 357 |
| G6PD | GCTGGAGTCCTG TCAGGCATTGC | TAGAGCTGAGGC GGAATGGGAG | 349 |
| HNF-4α | GCCTACCTCAAA GCCATCAT | GACCCTCCCAG CAGCATCTC | 256 |
| TAT | TGAGCAGTCTG TCCACTGCCT | ATGTGAATGAGG AGGATCTGAG | 338 |
| CYP2B6 | ATGGGGCACTG AAAAAGACTGA | AGAGGCGGGGA CACTGAATGAC | 283 |
| CYP2D6 | CTAAGGGAACGA CACTCATCAC | GTCACCAGGAA AGCAAAGACAC | 289 |
| CYP2C9 | CCTCTGGGGCA TTATCCATC | ATATTTGCACAGT GAAACATAGGA | 137 |
| CYP2C19 | TTCATGCCTTT CTCAGCAGG | ACAGATAGTGA AATTTGGAC | 277 |
| CYP2C8 | ACAACAAGCACCA CTCTGAGATATG | GTCTGCCAATTACA TGATCAATCTCT | 100 |
| CYP3A4 | GCCTGGTGCTC CTCTATCTA | GGCTGTTGACCA TCATAAAAGC | 187 |
| CYP1A1 | TCCAGAGACAA CAGGTAAAACA | AGGAAGGGCAG AGGAATGTGAT | 371 |
| CYP1A2 | ACCCCAGCTGC CCTACTTG | GCGTTGTGTC CCTTGTTGTG | 101 |
| CYP2E1 | ACCTGCCCCAT GAAGCAACC | GAAACAACTCC ATGCGAGCC | 246 |
| PXR | GAAGTCGGAG GTCCCCAAA | CTCCTGAAAAA GCCCTTGCA | 100 |
| CAR | TGATCAGCTGCA AGAGGAGA | AGGCCTAGCA ACTTCGCACA | 102 |

(continued)

| Gene | Forward primer | Reverse primer | Amplicon (bp) |
|------|----------------|----------------|---------------|
| AhR | ACATCACCTA CGCCAGTCGC | TCTATGCCGCT TGGAAGGAT | 101 |
| UGT1A1 | GGAGCAAAAGG CGCCATGGC | GTCCCCTCTG CTGCAGCTGC | 178 |
| LV-Bmi-1 | GCTGAGGGCTA TTGAGGCGCA | ACCCCAAATCCC CAGGAGCTGT | 127 |
| hBmi-1 | ACCTCCCAGCC CCGCAGAAT | AGACGCCGCTG TCAATGGGC | 280 |
| LV-hTERT | CAACCCGGCAC TGCCCTCAG | GGGGTTCCGCT GCCTGCAAA | 268 |
| hTERT | CGGAAGAGTGTC TGGAGCAAGT | GAACAGTGCCT TCACCCTCGA | 258 |

*PCR* Polymerase chain reaction, *ALB* albumin, *AFP* α-fetoprotein, *CK18* cytokeratin18, *G6PD* glucose-6-phosphate dehydrogenase, *HNF-4α* hepatocyte nuclear factor 4α, *TAT* tyrosine aminotransferase, *PXR* pregnane X receptor, *CAR* constitutive androstane receptor, *AhR* aryl hydrocarbon receptor, *UGT1A1* uridine diphosphate glucuronyltransferase, *LV-Bmi-1* lentivirus vector BMI-1, *hBmi-1* human Bmi-1, *LV-hTERT* lentivirus vector human telomerase reverse transcriptase, *hTERT* human telomerase reverse transcriptase

# References

1. Gomez-Lechon MJ, Donato MT, Castell JV, Jover R. Human hepatocytes in primary culture: the choice to investigate drug metabolism in man. Curr Drug Metab. 2004 Oct;5(5):443–62.

2. Rodriguez-Antona C, Donato MT, Boobis A, Edwards RJ, Watts PS, Castell JV, et al. Cytochrome P450 expression in human hepatocytes and hepatoma cell lines: molecular mechanisms that determine lower expression in cultured cells. Xenobiotica. 2002 Jun;32(6):505–20.

3. Gomez-Lechon MJ, Castell JV, Donato MT. Hepatocytes--the choice to investigate drug metabolism and toxicity in man: in vitro variability as a reflection of in vivo. Chem Biol Interact. 2007 May 20;168(1):30–50.

4. Guguen-Guillouzo C, Corlu A, Guillouzo A. Stem cell-derived hepatocytes and their use in toxicology. Toxicology. 2010 Mar 30; 270(1):3–9.

5. Khetani SR, Bhatia SN. Microscale culture of human liver cells for drug development. Nat Biotechnol. 2008 Jan;26(1):120–6.

6. Youdim KA, Tyman CA, Jones BC, Hyland R. Induction of cytochrome P450: assessment in an immortalized human hepatocyte cell line (Fa2N4) using a novel higher throughput cocktail assay. Drug Metab Dispos. 2007 Feb;35(2):275–82.

7. Sinz M, Wallace G, Sahi J. Current industrial practices in assessing CYP450 enzyme induction: preclinical and clinical. AAPS J. 2008 Jun;10(2):391–400.

8. Banas A, Yamamoto Y, Teratani T, Ochiya T. Stem cell plasticity: learning from hepatogenic differentiation strategies. Dev Dyn. 2007 Dec;236(12):3228–41.

9. Pittenger MF, Mackay AM, Beck SC, Jaiswal RK, Douglas R, Mosca JD, et al. Multilineage potential of adult human mesenchymal stem cells. Science. 1999 Apr 2;284(5411):143–7.

10. Jaiswal RK, Jaiswal N, Bruder SP, Mbalaviele G, Marshak DR, Pittenger MF. Adult human mesenchymal stem cell differentiation to the osteogenic or adipogenic lineage is regulated by mitogen-activated protein kinase. J Biol Chem. 2000 Mar 31;275(13):9645–52.

11. Nagai A, Kim WK, Lee HJ, Jeong HS, Kim KS, Hong SH, et al. Multilineage potential of stable human mesenchymal stem cell line derived from fetal marrow. PLoS One. 2007;2(12):e1272.

12. Ong SY, Dai H, Leong KW. Hepatic differentiation potential of commercially available human mesenchymal stem cells. Tissue Eng. 2006 Dec;12(12):3477–85.

13. Banas A, Teratani T, Yamamoto Y, Tokuhara M, Takeshita F, Quinn G, et al. Adipose

tissue-derived mesenchymal stem cells as a source of human hepatocytes. Hepatology. 2007 Jul;46(1):219–28.

14. Zemel R, Bachmetov L, Ad-El D, Abraham A, Tur-Kaspa R. Expression of liver-specific markers in naive adipose-derived mesenchymal stem cells. Liver Int. 2009 Oct;29(9):1326–37.

15. Yamamoto Y, Banas A, Murata S, Ishikawa M, Lim CR, Teratani T, et al. A comparative analysis of the transcriptome and signal pathways in hepatic differentiation of human adipose mesenchymal stem cells. FEBS J. 2008 Mar;275(6):1260–73.

16. Kulkarni JS, Khanna A. Functional hepatocyte-like cells derived from mouse embryonic stem cells: a novel in vitro hepatotoxicity model for drug screening. Toxicol In Vitro. 2006 Sep;20(6):1014–22.

17. Ek M, Soderdahl T, Kuppers-Munther B, Edsbagge J, Andersson TB, Bjorquist P, et al. Expression of drug metabolizing enzymes in hepatocyte-like cells derived from human embryonic stem cells. Biochem Pharmacol. 2007 Aug 1;74(3):496–503.

18. Unger C, Gao S, Cohen M, Jaconi M, Bergstrom R, Holm F, et al. Immortalized human skin fibroblast feeder cells support growth and maintenance of both human embryonic and induced pluripotent stem cells. Hum Reprod. 2009 Oct;24(10):2567–81.

19. Kutner RH, Zhang XY, Reiser J. Production, concentration and titration of pseudotyped HIV-1-based lentiviral vectors. Nat Protoc. 2009;4(4):495–505.

# Chapter 8

# Isolation of Adult Human Pluripotent Stem Cells from Mesenchymal Cell Populations and Their Application to Liver Damages

Shohei Wakao, Masaaki Kitada, Yasumasa Kuroda, and Mari Dezawa

## Abstract

We have found a novel type of pluripotent stem cells, Multilineage-differentiating stress enduring (Muse) cells that can be isolated from mesenchymal cell populations. Muse cells are characterized by stress tolerance, expression of pluripotency markers, self-renewal, and the ability to differentiate into endodermal-, mesodermal-, and ectodermal-lineage cells from a single cell, demonstrating that they are pluripotent stem cells. They can be isolated as cells positive for stage-specific embryonic antigen-3, a human pluripotent stem cell marker. Here, we introduce the isolation method for Muse cells and the effect of transplantation of these cells on chronic liver diseases.

**Key words:** Muse, Pluripotent stem cells, SSEA-3, Chronic liver diseases, Transplantation

## 1. Introduction

Recent progress of stem cell biology has demonstrated that application of stem cells to cell-based therapy is a realistic perspective toward the treatment of degenerative and traumatic diseases. It has been elucidated that cell therapy is also valid for the treatment of chronic liver diseases. As a cell source, it seems that mature hepatocytes are the most suitable for transplantation. However, it is still too difficult and insufficient to isolate hepatocytes from donor liver. Furthermore, in vitro long-term expansion of hepatocytes results in dysfunction of hepatocyte metabolism (1). Another candidate for a cell source toward transplantation to liver diseases is oval cells that have been proposed to be stem/progenitor cells in the liver (2, 3). While oval cells have an ability to differentiate into hepatocytes and biliary epithelial cells, they have been considered

Takahiro Ochiya (ed.), *Liver Stem Cells: Methods and Protocols*, Methods in Molecular Biology, vol. 826, DOI 10.1007/978-1-61779-468-1_8, © Springer Science+Business Media, LLC 2012

to appear only when the liver is damaged. Furthermore, their origin has not been elucidated yet. For these reasons, transplantation of oval cells is practically difficult. Thus, aiming for cell-based therapy toward chronic liver diseases, a practical cell source has been needed to be found (4).

Application of stem cells such as embryonic stem (ES) cells and bone marrow stromal cells to the damaged liver has been intensively studied since 1996 (5–8). ES cells have an ability to differentiate into hepatocytes, their clinical application has been tightly restricted because of the risk of tumorigenesity. On the contrary, bone marrow stromal cells have been already demonstrated to differentiate into hepatocytes in vitro and have been particularly successful in integration into the damaged liver (9, 10). Bone marrow stromal cells have been already applied to various types of clinical trials, and there have been obtained certain results in patients with liver diseases (11, 12).

While bone marrow stromal cells demonstrates differentiation into various kinds of cells, it is not clear what kind of cells assume the differentiation into hepatocytes and contribute to functional repair of damaged liver since bone marrow stromal cells are comprised of heterogeneous cell populations. We have previously found a unique type of stem cells, Muse cells, among adult human mesenchymal cell populations such as dermal skin fibroblasts and bone marrow stromal cells (13). Muse cells could be isolated as cells positive for stage-specific embryonic antigen-3 (SSEA-3), one of the human pluripotency markers, and generate a cell cluster from a single cell, whose appearance was very similar to that of human ES cell-derived embryoid bodies formed in suspension culture. Additionally, cells in a cluster derived from a single Muse cell were positive for the pluripotency markers such as Oct3/4, Sox2, and Nanog, and were able to spontaneously differentiate into endodermal- (hepatocytes and cholangiocytes), mesodermal- (smooth muscle cells and skeletal muscle cells), and ectodermal-lineage cells (epidermal cells and neural cells) when transferred onto the gelatin-coated dish. Muse cells were different from other pluripotent stem cells such as ES cells, in that, in spite of showing the ability to differentiate into cells representative of all three germ layers, they did not show tumorigenic proliferation activity, thus did not form teratomas when transplanted into immunodeficient mouse testes. Additionally, when human Muse cells were transplanted into immunodeficient mice with damaged liver caused by the intraperitoneal injection of $CCl_4$, they integrated into the damaged liver and expressed human albumin and human antitrypsin, indicating the differentiation of human Muse cells into functional hepatocytes in the mouse liver. These results suggest that Muse cells would be an advantageous source for cell-based therapy, since these cells are easily isolated from human general mesenchymal cells without showing tumorigenic proliferation activity.

## 2. Materials

### 2.1. Human Mesenchymal Cell Culture

1. Human mesenchymal cell source: Human mesenchymal cells are candidate cell source for isolation of Muse cells. We recommend to use adult human dermal fibroblasts (CC-2511, Lonza, Basel, Switzerland; 2320, ScienCell, Carlsbad, CA) and adult human bone marrow stromal cells (PT-2501, Lonza; ABM001, AllCells, Emeryville, CA) for isolation of Muse cells.

2. Culture medium for mesenchymal cells: alpha-minimum essential medium ($\alpha$-MEM) (M4526, Sigma-Aldrich, St Louis, MO) containing 10% fetal bovine serum (FBS) and 0.1 mg/ml kanamycin (15160054, Invitrogen, Carlsbad, CA). Store at 4°C.

### 2.2. Labeling of Muse Cells with Lentivirus-Mediated Gene Transfer

1. Packaging cell line: HEK293FT cells (R700-07, Invitrogen).

2. Expression vector and packaging plasmid: pWPXL, pMD2G and pCMV deltaR 8.74 provided by Dr Didier Trono, University of Geneva, Switzerland (14).

3. Transfection reagent: FuGENE HD Transfection Reagent (4709705, Roche Applied Science, Indianapolis, IN).

### 2.3. Isolation of Muse Cells by FACS Sorting

1. Fluorescence activated cell sorting (FACS) buffer: Calcium and magnesium-free 0.02 M phosphate-buffered saline (PBS) supplemented with 2 mM ethylenediaminetetraacetic acid (EDTA) and 0.5% bovine serum albumin (BSA) (01860–65, Nacalai tesque, Kyoto, Japan) with the filtration through a 0.22-µm Millex filter (SLGV033RS, Millipore, Billerica, MA). Store at 4°C (see Note 1).

2. Antibody: Primary antibody is anti-stage-specific embryonic antigen-3 (SSEA-3) antibody (1:50, MAB4303, Millipore), and secondary antibody is fluorescein isothiocyanate (FITC)-conjugated anti-rat IgM antibody (1:100, 112-095-075, Jackson ImmunoResearch, West Grove, PA) (see Note 2).

### 2.4. Suspension Culture for a Cluster Formation

1. Poly 2-hydroxyethyl methacrylate (poly-HEMA): 600 mg of poly-HEMA (P3932, Sigma-Aldrich) is dissolved in 40 ml of 95% ethanol gentry shaking at 37°C. Store at 4°C.

2. Methylcellulose: Methylcellulose (4100, MethoCult H4100, Stem Cell Technologies, Vancouver, BC) is diluted with 20% FBS in $\alpha$-MEM to a final concentration of 0.9%. Store at 4°C.

### 2.5. Immuno-cytochemistry

1. Paraformaldehyde (PFA) solution: 4 g of PFA is dissolved in about 50 ml of distilled water (DW) at 80°C with a few drops of 5 N sodium hydroxide solution. 25 ml of 0.4 M phosphate buffer (PB) is added to the solution. It is diluted to 100 ml with DW and filtered through a 0.80 µm Millex filter

(SLAA033SS, Millipore) to obtain 4% PFA solution dissolved in 0.1 M PB.

2. Blocking solution and antibody diluent: 0.02 M PBS supplemented with 5% normal goat serum (S-1000, Vector, Burlingame, CA), 0.3% BSA (A9418, Sigma-Aldrich), and 0.1% TritonX-100 (T8787, Sigma-Aldrich). Store at 4°C.

3. Antibody: Primary antibodies used are anti-Nanog (1:500, AB5731, Abcam, Cambridge, MA), anti-Oct3/4 (1:800, kindly provided by H. Hamada, Osaka University, Osaka, Japan), anti-Sox2 (1:1,000, AB5603, Abcam), and anti-SSEA-3 (Supernatant, 1:20, MC-631, Developmental Studies Hybridoma Bank, University of Iowa, Iowa city, Iowa). Secondary antibodies used are Alexa-488 or Alexa-568 conjugated anti-rabbit IgG or anti-mouse IgG antibodies (A11034, A11031, Molecular Probes, Invitrogen) and FITC-conjugated anti-rat IgM antibody (1:100).

**2.6. In Vitro Differentiation of Muse Cell**

1. Gelatin solution: 0.5 g of gelatin (G1890, Sigma-Aldrich) is dissolved in 500 ml of 0.02 M PBS followed by the autoclave sterilization to obtain 0.1% sterile gelatin solution. Store at 4°C.

2. Purification of total RNA: NucleoSpin RNA XS (740 902.10, Macherey-Nagel, Düren, Germany).

3. Antibody: Primary antibodies used are anti-$\alpha$ fetoprotein (1:10, N1501, Dako, Carpinteria, CA), anti-$\alpha$ neurofilament-M (1:200, AB1987, Abcam), and anti-$\alpha$-smooth muscle actin (MS-113-P1, Lab Vision, Fremont, CA). Secondary antibodies used are Alexa-488 or Alexa-568 conjugated anti-rabbit IgG or anti-mouse IgG antibodies (Molecular Probes, Invitrogen).

**2.7. CCl$_4$ Induced Liver Injury in Immuno-deficient Mice and Transplantation of Muse Cells**

1. Adult immunodeficient mice: 8–10-weeks-old male NOG mice (NOD/Shi-SCID, IL-2R$\gamma$KO Jic, ICLAS Monitoring Center, Kanagawa, Japan).

2. CCl$_4$: CCl$_4$ is diluted in olive oil to a final concentration of 10%.

**2.8. Immuno-histochemistry Analysis**

1. Periodate-lysine-paraformaldehyde (PLP) solution: 3.654 g of Lysin-HCl is dissolved in 100 ml of DW and 3.8 ml of 0.1 M Na$_2$HPO$_4$ (Solution A). 8 g of PFA is dissolved in about 50 ml of DW at 80°C with a few drops of 5 N sodium hydroxide solution, diluted to 100 ml with DW, and filtered through a 0.80-$\mu$m Millex filter (Millipore) to obtain 8% PFA in DW (Solution B). Solution A, Solution B and DW is mixed in a ratio of 15:2:3. Finally, 0.021 g of NaIO$_4$ is dissolved in 10 ml of the mixed solution (see Note 3).

2. Blocking solution and antibody diluents: 0.02 M PBS supplemented with 5% normal goat serum (Vector), 0.3% BSA (Sigma-Aldrich), and 0.1% TritonX-100 (Sigma-Aldrich). Store at 4°C.

3. Antibodies: Primary antibodies used are anti-human albumin (1:100, A80-229A , BETHYL Laboratories, Montgomery, TX), Golgi complex (1:200, AB6284, Abcam), and α1-anti-trypsin (1:200, RB-367-A1, Thermo Fisher Scientific, Waltham, MA). Secondary antibodies used are Alexa-568 conjugated anti-rabbit IgG and Alexa-680 conjugated anti-mouse IgG antibodies (A10042 and A21058, Molecular Probes, Invitrogen).

*2.9. Reverse Transcription-Polymerase Chain Reaction (RT-PCR)*

Following reagents are applied: TRIzol solution (15596–018, Invitrogen) for mRNA isolation, SuperScript II Reverse Transcriptase (18064–014, Invitrogen) for renerse transcription, and Ex Taq DNA polymerase (RR001, TaKaRa Bio, Shiga, Japan) for DNA amplification.

## 3. Method

*3.1. Preparation of Adult Human Mesenchymal Cells*

1. A frozen vial of mesenchymal cells is put into a 37°C water bath to thaw the frozen surface of the medium inside the vial. Do not completely thaw the medium.

2. The outside of the vial is sterilized with 70% ethanol.

3. Contents of the vial are transferred to a 15-ml conical tube.

4. 10 ml of the culture medium is added and the tube is centrifuged at $180 \times g$ for 5 min.

5. The supernatant is removed and the cells are resuspended in 10 ml of the culture medium.

6. The cells are transferred to a 100-mm culture dish and incubated at 37°C in a 5% $CO_2$ incubator.

7. When the cultured cells proliferate and reach 90% confluency, the cells are washed with 10 ml of PBS.

8. After removing PBS, 2 ml of 0.25% trypsin solution (25200, Invitrogen) is added and the cells are incubated at 37°C for 3 min.

9. 8 ml of the culture medium is added and the cells are transferred to a 15-ml conical tube.

10. The cells are centrifuged at $210 \times g$ for 3 min.

11. After removing the supernatant, the cells are seeded to two new 100-mm culture dishes.

*3.2. Labeling of Muse Cells with Lentivirus-Mediated Gene Transfer*

1. At day 0, HEK293FT cells are to be 50~70% confluent in a 100-mm culture dish and are washed with 10 ml of PBS.

2. After removing PBS, 2 ml of 0.25% trypsin solution is added and the cells are incubated at 37°C for 3 min.

3. 8 ml of the culture medium is added and the cells are transferred to a 15-ml conical tube.

4. The cells are centrifuged at $210 \times g$ for 3 min.

5. After removing the supernatant, the concentration of the cells is adjusted to $5 \times 10^5$ cells per ml with the culture medium without antibiotics.

6. At day 1, in a 1.5-ml tube, 10 μg of each plasmid DNA such as pWPXL, pMD2G, and pCMV deltaR 8.74 is mixed in 500 μl of Opti-MEM I medium.

7. After 90 μl of FuGENE HD Transfection Reagent is added, the mixture of the transfection complex is briefly pipetted and incubated at room temperature for 15 min.

8. The mixture of the transfection complex is applied in a dropwise manner to the dish, in which HEK293FT cells are cultured, and the cells are incubated at 37°C in a 5% $CO_2$ incubator.

9. At day 2, the culture medium containing the transfection complex is removed and the new culture medium is applied.

10. At day 4, the virus-containing supernatant is harvested and transferred to a 15-ml conical tube.

11. The supernatant is centrifuged at $730 \times g$ for 15 min at 4°C to pellet the cell debris.

12. The supernatant is filtered through a sterile, 0.45-μm low protein-binding filter.

13. The supernatant is applied to the 100 mm culture dish in which $1 \times 10^6$ adult human mesenchymal cells is cultured to label the cells by lentivirus encoding GFP.

*3.3. Isolation of Muse Cells from Adult Human Mesenchymal Cells*

1. When the cells are reached 100% confluency, the culture medium is removed, and the cells are washed with 10 ml of PBS.

2. After removing PBS, the cells are detached by incubation in 2 ml of 0.25% trypsin solution at 37°C for 5 min (see Note 4).

3. The detached cells are suspended in the trypsin solution supplemented with 9 ml of the culture medium and transferred to a 15-ml conical tube.

4. The cells are centrifuged at $210 \times g$ for 5 min.

5. The supernatant is removed and the cells are resuspended in 10 ml of PBS.

6. The cells are centrifuged at $210 \times g$ for 5 min.

7. The supernatant is removed and the cells are resuspended with 10 ml of the FACS buffer.

8. The cells are centrifuged at $210 \times g$ for 5 min.

9. The supernatant is removed and the concentration of the cells is adjusted to $1 \times 10^6$ cells per 100 μl of the FACS buffer.

10. 2 μl of anti-SSEA-3 antibody is added.

11. The cells are incubated on ice for 1 h with brief and gentle tapping every 10 min (see Note 5).

12. The cells are centrifuged at $400 \times g$ for 5 min at 4°C.

13. The supernatant is removed and the cells are resuspended in 1 ml of the FACS buffer (see Note 6).

14. Repeat the processes in 12 and 13, two times.

15. The cells are centrifuged at $400 \times g$ for 5 min at 4°C.

16. The supernatant is removed and the concentration of the cells is adjusted to $1 \times 10^6$ cells per 100 μl of FACS buffer.

17. 1 μl of FITC-conjugated anti-rat IgM antibody is added (see Notes 7 and 8).

18. The cells are incubated on ice for 1 h with brief and gentle tapping every 10 min.

19. The cells are centrifuged at $400 \times g$ for 5 min at 4°C.

19. The cells are centrifuged at $400 \times g$ for 5 min at 4°C.

20. The supernatant is removed and the cells are resuspended in 1 ml of the FACS buffer.

21. Repeat the processes in 18 and 19, two times.

22. The cells are centrifuged at $400 \times g$ for 5 min at 4°C.

23. The supernatant is removed and the concentration of the cells is adjusted to $1 \times 10^6$ cells per 200 μl of FACS buffer.

24. SSEA-3 (+) cells are isolated by SORP FACSAria II (Beckton Dickinson, Franklin Lakes, NJ) (Fig. 1) (see Notes 9 and 10).

**3.4. Suspension Culture for a Cluster Formation**

1. Prior to the culturing Muse cells, the culture dish is coated with poly-HEMA solution to avoid the adhesion of Muse cells to the bottom of the dish (Table 1).

2. The dish is air-dried overnight in the clean bench.

3. Methylcellulose and Muse cells are mixed thoroughly by gentle pipetting, and the mixture is transferred onto a poly-HEMA-coated dish (Table 2).

4. To prevent drying, 1/10 volume of 15% FBS in α-MEM is gently added to the dish every 3 days.

**3.5. Immuno-cytochemistry**

1. Collected Muse cell-derived clusters (M-Clusters) are fixed with PFA solution.

2. The M-clusters are centrifuged at $100 \times g$ for 5 min, embedded with OCT compound using dry ice, and cut into 8 μm-thick cryosections.

3. The sections are dried for 10 min at room temperature.

4. The sections are washed with 0.02 M PBS three times and incubated with the blocking solution at room temperature for 30 min.

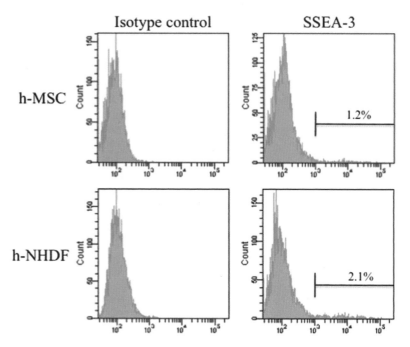

Fig. 1. Flow cytometry analysis of SSEA-3-positive Muse cells in human bone marrow stromal cells (h-MSC) and adult human dermal fibroblasts (h-NHDF). H-MSC and h-NHDF contain approximately 1 and 2% SSEA-3-positive Muse cells, respectively.

## Table 1
## Dosage recommendation of poly-HEMA solution

| Dish | Poly-HEMA |
| --- | --- |
| 100 mm | 6.5 ml |
| 60 mm | 2.5 ml |
| 35 mm | 1 ml |
| 12 well | 400 µl |
| 24 well | 200 µl |
| 48 well | 130 µl |
| 96 well | 40 µl |

The culture dishes for suspension culture were coated with poly-HEMA to avoid the adhesion of Muse cells to the bottom of the dish. The amount of poly-HEMA solution is dependent on the size of the dish

**Table 2**
**Dosage recommendation of methylcellulose culture medium**

| Dish | Cell Number | Cell + αMEM (µl) | FBS | 2.6% MC (µl) | Total (µl) |
|------|-------------|------------------|-----|--------------|------------|
| 100 mm | $1.6 \times 10^5$ | 9,920 | 2,880 | 6,400 | 19,200 |
| 60 mm | $6 \times 10^4$ | 3,720 | 1,080 | 2,400 | 7,200 |
| 35 mm | $2.5 \times 10^4$ | 1,550 | 450 | 1,000 | 3,000 |
| 12 well | $1 \times 10^4$ | 642.5 | 187.5 | 420 | 1,250 |
| 24 well | $5 \times 10^3$ | 365 | 105 | 230 | 700 |
| 48 well | $3 \times 10^3$ | 210 | 60 | 130 | 400 |
| 96 well | $1 \times 10^3$ | 70.5 | 19.5 | 40 | 130 |

The amount of Methyl cellulose culture medium is dependent on the size of the dish. It should be applied to the poly-HEMA-coated dish

5. The sections are then incubated with the primary antibodies in the blocking solution at 4°C for overnight.

6. The sections are washed with 0.02 M PBS three times and incubated with the secondary antibodies and 4′,6′-diamidino-2-phenylindole (DAPI) in the blocking solution for 2 h at room temperature.

7. The sections are washed with 0.02 M PBS three times and enclosed with the cover glass (Fig. 2a–g).

*3.6. In Vitro Differentiation of a M-Cluster*

1. The culture dish is coated with 0.1% gelatin solution.

2. The dish is incubated at 37°C for 30 min and the gelatin solution is removed just prior to use. Avoid drying.

3. At day 7–10 in methylcellulose culturing, the equal volume of α-MEM is added, and the mixture of the methylcellulose and α-MEM is gently pipetted.

4. Methylcellulose is transferred to a 15-ml conical tube and the culture medium is added to reach 14 ml.

5. The cells are centrifuged at $210 \times g$ for 15 min.

6. The supernatant is removed to leave 2 ml of the medium and 12 ml of α-MEM is added.

7. The cells are centrifuged at $210 \times g$ for 10 min.

8. The supernatant is removed to leave 2 ml of the medium and 9 ml of α-MEM is added.

9. The cells are centrifuged at $100 \times g$ for 5 min.

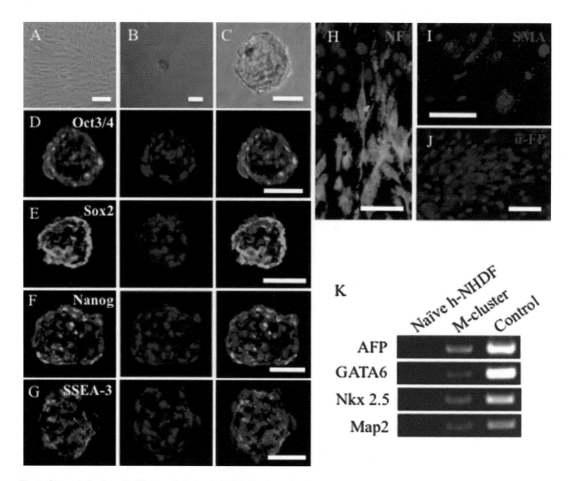

Fig. 2. Characterization of a Muse cell-derived cluster (M-cluster), which is formed from a single Muse cell, and differentiation of cells derived from the M-cluster. (a) Adherent culture of adult human dermal fibroblasts (h-NHDF). (b, c) Methylcellulose culture of Muse cells derived from h-NHDF on day 7 (b: low magnification, (c) high magnification). Immunocytochemical localization of Oct3/4 (d), Sox2 (e), Nanog (f) and SSEA-3 (g) in M-clusters derived from h-NHDF. Immunocytochemistry of Neurofilament-M (NF) (h), α smooth muscle actin (SMA) (i), and α-fetoprotein (AFP) (j), in cells derived from a single M-cluster. (K) RT-PCR analysis of naïve h-NHDF and M-cluster derived from h-NHDF. Whole human fetus is used as the positive control. (Scale bars: 50 μm).

10. The supernatant is removed to reach 2 ml remaining inside the tube and 8 ml of α-MEM is added.

11. The suspension M-clusters are transferred to a 100-mm dish.

12. The M-clusters are picked up with a glass micropipette under the phase-contrast microscope and transferred onto the gelatin-coated dish containing 10% FBS in α-MEM.

13. At day 7 in culturing on a gelatin-coated dish, total RNA is extracted from the expanded cells and purified using NucleoSpin RNA XS. Another expanded cells are fixed with PFA solution and performed immunostaining with the endodermal, mesodermal, and ectodermal markers (Fig. 2h–k).

### 3.7. CCl₄ Induced Liver Injury in Immunodeficient Mice and Transplantation of Muse Cells

1. At day 0, immunodeficient mice undergo intraperitoneal injection of 100 μl per 20 g body weight of olive oil containing 10% $CCl_4$.

2. At day 1, Muse cells are collected by FACS, and the concentration of the cells is adjusted to $1 \times 10^5$ cells per 100 μl of saline.

3. Avertin solution is administered to the mouse by intraperitoneal injection for anesthesia (see Note 11).

4. Muse cells are slowly transplanted by tail vein injection with a 30 G-needle that is for dental use.

5. After transplantation, the mouse is warmed prior to placing back to the cage after waking from anesthesia.

### 3.8. Immunohistochemistry Analysis

1. At 4 weeks after transplantation, the mice are fixed with PLP solution.

2. The livers are harvested and soaked into the same fixatives for 6 h at 4°C.

3. Tissues are then washed with 0.02 M PBS and incubated in 15, 20, and 25% sucrose in 0.02 M PBS for 12 h each at 4°C.

4. Tissues are embedded with OCT compound using dry ice and cut into 10-μm thick cryosections.

5. The sections are dried for 10 min at room temperature.

6. The sections are washed with 0.02 M PBS and incubated with the blocking solution at room temperature for 30 min.

7. The sections are incubated with the primary antibodies in the blocking solution at 4°C for overnight.

8. The sections are washed with 0.02 M PBS at three times and incubated with the secondary antibodies and DAPI in the blocking solution for 2 h at room temperature.

9. The sections are washed with 0.02 M PBS at three times and enclosed with the cover glass (Fig. 3a, b).

### 3.9. Reverse Transcription-Polymerase Chain Reaction (RT-PCR)

1. After 4 weeks since transplantation, the livers are harvested and washed with 0.02 M PBS.

2. Tissues are immersed in TRIzol solution and homogenized according to the manufacturer's protocol.

3. The concentration of collected total RNA is measured by NanoDrop 1000 (Thermo Fisher Scientific).

4. 0.5 μg of total RNA is reverse-transcribed with SuperScript II Reverse Transcriptase according to the manufacturer's protocol.

5. PCR condition: an initial denaturing at 94°C for 5 min and 35 cycles of denaturing step at 94°C for 1 min, annealing step at 60°C for 1 min, and an extension step at 72°C 1 min (Fig. 3c).

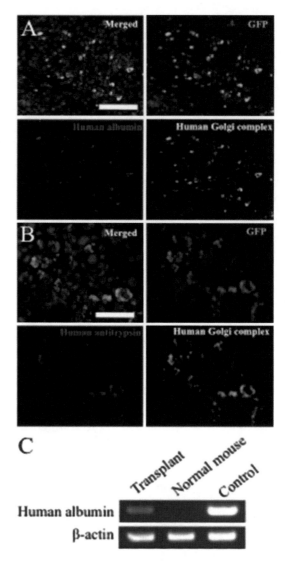

Fig. 3. Differentiation of GFP-labeled transplanted Muse cells in the damaged liver of immunodeficient mice. At 4 weeks after transplantation, transplanted GFP-positive Muse cells in the damaged liver are positive for human Golgi complex, and some of these cells are positive for human albumin (**a**) or human α1-antitrypsin (**b**). (**c**) Gene expression of human albumin in the damaged liver analyzed by RT-PCR. RNA is isolated from the livers of Muse cell-transplanted or normal mice. Human fetus liver is used as the positive control. (Scale bars: A, 50μm, B, 100 μm).

## 4. Notes

1. Because adult human mesenchymal cells are highly adherent, the FACS buffer should contain 2 mM EDTA.

2. Anti-SSEA-3 antibody used in this study is purchased from Millipore. Any other commercially available SSEA-3 antibodies are not recommended.

3. PLP solution should be freshly prepared just before use.

4. Be sure to confirm that the cells are dissociated into single cells by trypsin treatment.

5. Be careful for tapping. Too much tapping results in the increase of the dead cells that will increase viscosity of the cell suspension. If more than 500 µl of cell suspension are subjected to staining, mixing of cells should be done by inverting.

6. 100 µl of the supernatant should remain after removing it, and then the FACS buffer is added after stirring with the micro tube mixer (MT-360, Tomy Seiko, Tokyo, Japan) for a few seconds.

7. The secondary antibody should be centrifuged at $15,000 \times g$ for 2 min at 4°C before use to avoid contamination of the aggregated fluorescent dyes.

8. If the cells are labeled with the lentivirus encording GFP, the DyLight649 conjugated secondary antibody can be applied instead of that conjugated to FITC.

9. Flow cytometry with anti-SSEA-3 antibody gives the linearly continuous pattern in which the positive and negative fractions do not clearly separate. For this, both the isotype control and negative control in which only the secondary antibody is applied should be used for determining the gate of the SSEA-3-positive fraction.

10. The SSEA-3-positive population varies depending on the confluency and number of passages. For FACS sorting of SSEA-3-positive fraction, cultured mesenchymal cells should be in 100% confluency, and the number of passages of these cells should be less than 11.

11. Dosage recommendation of avertin is weight (g) × 0.02 µl + 0.05 µl.

## References

1. Beigel J., Fella K., Kramer PJ., Kroeger M., Hewitt P. (2008) Genomics and proteomics analysis of cultured primary rat hepatocytes. *Toxicol In Vitro* **22**, 171–181.

2. Lemire JM., Shiojiri N., Fausto N. (1991) Oval cell proliferation and the origin of small hepatocytes in liver injury induced by D-galactosamine. *Am J Pathol* **139**, 535–552.

3. Wang X., Foster M., Al-Dhalimy M., Lagasse E., Finegold M., Grompe M. (2003) The origin and liver repopulating capacity of murine oval cell. *Proc Natl Acad Sci USA* 100 Suppl 1:11881–11888.

4. Seglen P., Sacter G., Schwarze P. (1994) Proliferation of diploid hepatocytes and nonparenchymal (oval) cells during rat-liver regeneration in the presence of 2-acetylaminofluorene. *Int J Oncol.* 5(4):805–810.

5. Abe K., Niwa H., Iwase K., Takiguchi M., Mori M., Abe SI., Abe K., Yamamura KI. (1996) Endoderm-specific gene expression in embryonic stem cells differentiated to embryoid bodies. *Exp Cell Res* **229**, 27–34.

6. Teratani T., Yamamoto H., Aoyagi K., Sasaki H., Asari A., Quinn G., Sasaki H., Terada M., Ochiya T. (2005) Direct hepatic fate specification from mouse embryonic stem cells. *Hepatology* **41**, 836–846.

7. Cai J., Zhao Y., Liu Y., Song Z., Qin H., Meng S., Chen Y., Zhou R., Song X., Guo Y., Ding M.,

Deng H. (2007) Directed differentiation of human embryonic stem cells into functional hepatic cells. *Hepatology* **45**, 1229–1239

8. Hay DC., Fletcher J., Payne C., Terrace JD., Gallagher RC., Snoeys J., Black JR., Wojtacha D., Samuel K., Hannoun Z., Pryde A., Filippi C., Currie IS., Forbes SJ., Ross A., Newsome PN., Iredale JP. (2008) Highly efficient differentiation of hESCs to functional hepatic endoderm requires ActivinA and Wnt3a signaling. *Proc Natl Acad Sci USA* **105**, 12301–12306.

9. Okumoto K., Saito T., Hattori E., Ito JI., Adachi T., Takeda T., Sugahara K., Watanabe H., Saito K., Togashi H., Kawata S. (2003) Differentiation of bone marrow cells into cells that express liver-specific genes in vitro: implication of the Notch signals in differentiation. *Biochem Biophys Res Commun* **304**, 691–695.

10. Oyagi S., Hirose M., Kojima M., Okuyama M., Kawase M., Nakamura T., Ohgushi H., Yagi K. (2006) Therapeutic effect of transplanting HGF-treated bone marrow mesenchymal cells into CCl4-injured rats. *J Hepatol* **44**, 742–748.

11. Theise ND., Nimmakayalu M., Gardner R., Illei PB., Morgan G., Teperman L., Henegariu O., Krause DS. (2000) Liver from bone marrow in humans. *Hepatology* **32**, 11–16.

12. Terai S., Ishikawa T., Omori K., Aoyama K., Marumoto Y., Urata Y., Tokoyama Y., Uchida K., Yamasaki T., Fujii Y., Okita K., Sakaida I. (2006) Improved liver function in patients with liver cirrhosis after autologous bone marrow cell infusion therapy. *Stem Cells* **24**, 2292–2298.

13. Kuroda Y., Kitada M., Wakao S., Nishikawa K., Tanimura Y., Makinoshima H., Goda M., Akashi H., Inutsuka A., Niwa A., Shigemoto T., Nabeshima Y., Nakahata T., Nabeshima Y., Fujiyoshi Y., Dezawa M. (2010) Unique multipotent cells in adult human mesenchymal cell populations. *Proc Natl Acad Sci USA* **107**, 8639–8643.

14. Nguyen TH., Khakhoulina T., Simmons A., Morel P., Trono D. (2005) A simple and highly effective method for the stable transduction of uncultured porcine hepatocytes using lentiviral vector. *Cell Transplant* **14**, 489-496.

# Chapter 9

# Generation and Hepatic Differentiation of Human iPS Cells

## Tetsuya Ishikawa, Keitaro Hagiwara, and Takahiro Ochiya

## Abstract

A method for the generation of human induced pluripotent stem (iPS) cells was established. This method employs adenovirus carrying the ecotropic retrovirus receptor mCAT1 and Moloney murine leukemia virus (MMLV)-based retroviral vectors carrying the four transcription factors POU5F1 (OCT3/4), KLF4, SOX2, and MYC (c-Myc) (Masaki H & Ishikawa T Stem Cell Res 1:105–15, 2007). The differentiation of human iPS cells into hepatic cells was performed by a stepwise protocol (Song Z et al. Cell Res 19:1233–42, 2009). These cells have potential as patient-specific in vitro models for studying disease etiology and could be used in drug discovery programs tailored to deal with genetic variations in drug efficacy and toxicity.

**Key words:** Human iPS cells, POU5F1 (OCT3/4), KLF4, SOX2, MYC (c-Myc), Retrovirus, Hepatic differentiation

## 1. Introduction

Human pluripotent stem cells can be established from the postnatal tissue of a patient and it would greatly facilitate the development of patient-specific in vitro models for studying disease etiology and help understand the role of genetic variation in drug responses. We applied a method for the transduction of human neonatal dermal cells. This method employs adenovirus carrying the ecotropic retrovirus receptor mCAT1 and Moloney murine leukemia virus (MMLV)- based retroviral vectors carrying the four transcription factors POU5F1 (OCT3/4), KLF4, SOX2, and MYC (c-Myc) (1). Among the small cells induced by the four-gene transduction, only colonies with defined edges were established by repeated cloning under conditions of human embryonic stem (ES) cell culture. Subsequently established human induced pluripotent stem (iPS) cells were similar to human ES cells in morphology, alkaline phosphatase activity, cell surface markers, gene expression, DNA methylation of the promoter region of OCT3/4 and NANOG, long-term

Takahiro Ochiya (ed.), *Liver Stem Cells: Methods and Protocols*, Methods in Molecular Biology, vol. 826,
DOI 10.1007/978-1-61779-468-1_9, © Springer Science+Business Media, LLC 2012

self-renewal ability, and teratoma formation. The four transgenes were silenced in iPS cells. Genome-wide single-nucleotide polymorphism array analysis revealed no marked differences between human iPS cells and their parental cells. Consistent with these observations, the HLA genotypes for the human iPS cells and the parental cells were identical.

The hepatic differentiation of human iPS cells has been described (2). In this report, human iPS cells were induced to differentiate into hepatic cells by a stepwise protocol. Approximately 60% of the differentiated human iPS cells at day 7 expressed the hepatic markers alpha fetoprotein and albumin. The differentiated cells at day 21 exhibited liver cell functions, including albumin secretion, glycogen synthesis, urea production, and inducible cytochrome P450 activity.

## 2. Materials

### 2.1. Viral Infection of Human Neonatal Dermal Cells

1. Adenovirus vector plasmids (Takara Bio Clontech).

2. HEK293 cells (MicroBix).

3. Adenovirus purification kit (Takara Bio Clontech).

4. The Adeno-X rapid titer kit for titer determination for vector stocks (Takara Bio Clontech).

5. pMXs retrovirus vector: The replication-incompetent MMLV-based retrovirus vector (3).

6. Recombinant retroviruses: Recombinant retroviruses were generated by transfection of vectors into the Plat-E packaging cells (3), followed by incubation in FBM medium (obtained by Lonza, Baltimore, MD, USA) supplemented with FGM-2 SingleQuots (obtained by Lonza, Baltimore, MD, USA).

7. Mouse embryonic fibroblast-conditioned medium (MEF-CM): DMEM/F12 (Gibco) supplemented with 20% Knockout Serum Replacement (Invitrogen), 2 mM L-glutamine (Sigma), 1× nonessential amino acids (Sigma), 10 μg/ml gentamycin (Gibco), and 10 ng/ml bFGF (Peprotech) was conditioned on mitomycin C-treated mouse embryonic fibroblasts (MEFs) (Reprocell) for 20–24 h, harvested, filtered through a 0.45-μm filter, and supplemented with 0.1 mM 2-mercaptoethanol (Sigma) and 10 ng/ml bFGF before use.

8. Fibroblasts: Fibroblasts were obtained from human neonatal foreskin, by biopsy under informed consent (Lonza, Baltimore, MD, USA) followed by culture in FBM supplemented with FGM-2 SingleQuots.

9. Retrovirus/Polybrene solution: Equal volumes of the retrovirus vector suspension for each of the four genes being considered

(OCT3/4, SOX2, MYC (c-Myc), and KLF4) were supplemented with 5 µg/ml Polybrene.

**2.2. Maintenance of Human iPS Cells**

1. Human iPS cell media and matrices: DMEM/F12 (Gibco) supplemented with 20% Knockout Serum Replacement (Invitrogen), 2 mM L-glutamine (Sigma), 1× nonessential amino acids (Sigma), 10 µg/ml gentamycin (Gibco), 10 ng/ml bFGF (Peprotech), and 0.1 mM 2-mercaptoethanol. MEFs (Reprocell) . Matrigel (BD Biosciences). mTeSR1 medium (Stem Cell Technologies).

2. 10 mM ROCK inhibitor (Y-27632, Calbiochem) solution.

3. Hanks' balanced salt solution (Gibco).

4. 0.25% trypsin–EDTA (Gibco).

5. Cell Freezing Solution for hES cells (Reprocell).

**2.3. Karyotype Analysis and Fluorescence In Situ Hybridization**

1. 0.02 µg/ml Colcemid (Nacalai).

2. 0.06–0.075 M KCl.

3. Carnoy's fixative.

4. Multicolor FISH probe (Cambio).

5. Fluorescence microcopy (Axio Imager Z1, Zeiss).

**2.4. Real-Time PCR and DNA Microarray Analyses**

1. RecoverAll Total Nucleic Acid Isolation kit (Ambion).

2. RNeasy (Qiagen).

3. TaqMan preamp (Applied Biosystems).

4. ABI Prism 7900HT Thermocycler (Applied Biosystems).

5. Human Genome U133 Plus 2.0 gene expression arrays (Affymetrix) or Whole Human Genome Oligo microarray (Agilent).

6. GeneSpring GX 10.0 software (Agilent).

7. Primers for RT-PCR gene expression analysis are listed in Tables 1 and 2.

**2.5. Alkaline Phosphatase Staining and Immuno-cytochemistry**

1. 10% neutral-buffered formalin solution (Wako).

2. Phosphate-buffered saline (PBS).

3. Alkaline phosphatase detection kit (One-Step NBT/BCIP; Pierce).

4. 10% formaldehyde.

5. 0.1% gelatin/PBS.

6. Primary antibodies against SSEA-3 (MC-631; Chemicon), SSEA-4 (MC813-70; Chemicon), TRA-1-60 (ab16288; abcam), TRA-1-81 (ab16289; abcam), CD9 (M-L13; R&D systems), CD24 (ALB9; abcam), Thy1 (5E10; BD Bioscience), or Nanog (MAB1997; R&D Systems).

7. 0.1% Triton X-100/PBS solution.

## Table 1
## PCR primers used for gene expression analyses

| Target | Forward primer sequence | Reverse primer sequence |
| --- | --- | --- |
| HPRT1 | AGTCTGGCTTATATCCAACACTTCG | GACTTTGCTTTCCTTGGTCAGG |
| NANOG | TACCTCAGCCTCCAGCAGAT | TGCGTCACACCATTGCTATT |
| TERT | AGCCAGTCTCACCTTCAACCGC | GGAGTAGCAGAGGGAGGCCG |
| SALL4 | AAACCCCAGCACATCAACTC | GTCATTCCCTGGGTGGTTC |
| ZFP42 | TTGGAGTGCAATGGTGTGAT | TCTGTTCACACAGGCTCCAG |
| GDF3 | GGCGTCCGCGGGAATGTACTTC | TGGCTTAGGGGTGGTCTGGCC |
| DNMT3B | GCAGCGACCAGTCCTCCGACT | AACGTGGGGAAGGCCTGTGC |
| TDGF1 | ACAGAACCTGCTGCCTGAAT | AGAAATGCCTGAGGAAAGCA |
| GABRB3 | CTTGACAATCGAGTGGCTGA | TCATCCGTGGTGTAGCCATA |
| CYP26A1 | AACCTGCACGACTCCTCGCACA | AGGATGCGCATGGCGATTCG |

## Table 2
## PCR primers used for gene expression analyses

| Target | Forward primer sequence | Reverse primer sequence |
| --- | --- | --- |
| OCT3/4-total | GAGAAGGAGAAGCTGGAGCA | AATAGAACCCCCAGGGTGAG |
| OCT3/4-exo | AGTAGACGGCATCGCAGCTTGG | GGAAGCTTAGCCAGGTCCGAGG |
| SOX2-total | CAGGAGAACCCCAAGATGC | GCAGCCGCTTAGCCTCG |
| SOX2-exo | ACACTGCCCCTCTCACACAT | CGGGACTATGGTTGCTGACT |
| KLF4-total | ACCCTGGGTCTTGAGGAAGT | ACGATCGTCTTCCCCTCTTT |
| KLF4-exo | CTCACCCTTACCGAGTCGGCG | GCAGCTGGGGCACCTGAACC |
| c-Myc-total | TCCAGCTTGTACCTGCA GGATCTGA | CCTCCAGCAGAAGGTGATCCAGACT |
| c-Myc-exo | AGTAGACGGCATCGCAGCTTGG | CCTCCAGCAGAAGGTGATCCAGACT |

8. AlexaFluor 488-conjugated secondary antibodies (Molecular Probes).

9. Hoechst 33258 (Nacalai).

10. Axiovert 200 M microscope (Carl Zeiss).

***2.6. Histology and Immuno-histochemistry***

1. Mounting medium for embedding teratomas.

2. Hematoxylin–eosin (HE) staining solution.

3. Alcian blue staining solution.

4. Immunoblock (Dainippon–Sumitomo).

5. Primary antibodies: anti-nestin polyclonal antibody (PRB-570C, Covance, 1:300), anti-type II collagen polyclonal antibody (LB-1297, LSL, 1:200), anti-smooth muscle actin polyclonal antibody (RB-9010-R7, Lab Vision, 1:1), anti-$\alpha$-fetoprotein polyclonal antibody (A0008, Dako, 1:500), anti-MUC-1 polyclonal antibody (RB-9222-P0, Lab Vision, 1:100), and anti-human nuclei monoclonal antibody (HuNu) (MAB1281, Chemicon, 1:300).

6. Hyaluronidase.

7. Secondary antibodies (Alexa Fluor 594 and 688, Molecular Probes, 1:600).

8. DAPI (InnoGenex).

9. Fluorescence microscope: Axio Imager Z1 (Zeiss).

*2.7. Methylation Analysis*

1. Primers containing a T7 promoter (Table 3).

2. RNase A.

3. Epityper for MALDI-TOF mass spectrometry (Sequenom).

4. MassARRAY mass spectrometer (Bruker Sequenom).

## Table 3
## Amplicons and primer sequences for methylation analysis

| Amplicon name | Location in genome | Size of amplicon | Name of primers | Sequences of primers |
|---|---|---|---|---|
| OCT3/4-z1 | chr6:31248581 -31249029 | 449 | OCT3/4-z1-L | aggaagagagTAGTAGGGATT TTTTGGA TTGGTTT |
| | | | OCT3M-z1-R | cagtaatacgactcactatagggagaaggctAAAA CTTTTCCCCCACTCTTATATTAC |
| OCT3/4-z2 | chr6_qbl_ hap2:23882 99-2388525 | 227 | OCT3/4-z2-L | aggaagagagGGTAATAAAGTG AGATTT TGTTTTAAAAA |
| | | | OCT3/4-z2-R | cagtaatacgactcactatagggagaaggctCCA CCCACTAACCTTAACCTCTAA |
| NANOG-z1 | chr12:7832645- 7832959 | 315 | NANOG-z1-L | aggaagagagGGAATTTAAGGTGT ATGT ATTTTTTATTTT |
| | | | NANOG-z1-R | cagtaatacgactcactatagggagaaggctATAA CCCACCCCTATAATCCCAATA |
| NANOG-z2 | chr12:7832877- 7833269 | 393 | NANOG-z2-L | aggaagagagGTTAGGTTGGTT TTAAAT TTTTGAT |
| | | | NANOG-z2-R | cagtaatacgactcactatagggagaaggctTTT ATAATAAAAACTCTATCA CCTTAAACC |

| | |
|---|---|
| **2.8. Southern Blot Analysis and Genomic PCR** | 1. 0.8% agarose gel. |
| | 2. QIAquick gel extraction kit (QIAGEN). |
| | 3. $(\gamma-{}^{32}P)$ATP. |
| | 4. *Kpn*I restriction enzyme(Takara Bio). |
| | 5. HybondXL membrane (GE Healthcare). |

**2.9. SNP Genotyping and HLA Typing**

1. GeneChip Human Mapping 500 K Array Set (Affymetrix).
2. *Nsp*I and *Sty*I restriction enzymes.
3. Biotin labeled primers.
4. Color-coded microbeads. Hybridization of amplified DNA was identified by cytometry dual-laser analysis.

**2.10. Culture and Differentiation of Human iPS Cells into Definitive Endoderm and Hepatocyte**

1. RPMI 1640 medium (Invitrogen/Gibco, Rockville, MD, USA), supplemented with 0.5 mg/ml albumin fraction V (Sigma-Aldrich, St Louis, MO, USA) and 100 ng/ml activin A (Peprotech).
2. 0.1 and 1% insulin-transferrin-selenium (Invitrogen).
3. Activin A (Peprotech).
4. Hepatocyte Culture Medium (HCM) (Lonza, Baltimore, MD, USA).
5. FGF4, BMP2, HGF, KGF (Peprotech).
6. Oncostatin-M (R&D, Minneapolis, MN, USA).
7. Dexamethasone (Sigma-Aldrich).
8. Dulbecco's Modified Eagle's Medium (DMEM) containing N2, B27, 1 mM/l GlutaMAX™, 1% nonessential amino acids, and 0.1 mM β-mercaptoethanol (all from Invitrogen/Gibco).

**2.11. Immuno-fluorescence Assay**

1. 4% paraformaldehyde phosphate buffer solution.
2. Goat and rabbit serum.
3. 0.1% TritonX-100 (Fisher, UK) in PBS.
4. Primary antibodies: human rabbit anti OCT3/4 (Abcam, La Jolla, CA, USA), goat anti NANOG (R&D), mouse anti SSEA4, mouse anti TRA-1-60, mouse anti TRA-1-81 (Santa Cruz, CA), rabbit anti Ki67 (Invitrogen), goat anti SOX17 (R&D), rabbit anti FOXA2 (Upstate), mouse anti CK18 and mouse anti AFP (Invitrogen), rabbit anti ALB (DAKO, Glostrup, Denmark), rabbit anti AAT (Invitrogen) and rabbit anti CYP3A4 (AbD Serotec, Oxford, UK).
5. FITC-conjugated or TRITC-conjugated secondary antibodies (Invitrogen).
6. 4,6-Diamidino-2-phenylindole (Roche, Germany).

### 2.12. PAS Stain for Glycogen

1. PAS staining system (Sigma-Aldrich).
2. Amylase (Sigma).

### 2.13. Urea Nitrogen Kinetic Quantitative Determination and Albumin Secretion ELISA Assay

1. Urea nitrogen determination system (STANBIO, Boerne, TX, USA).
2. 0.25% Trypsin (Gibco).
3. Hemocytometer.
4. Synergy HT Multi-Detection Microplate Reader (BioTek) with Gen5 software (BioTek).
5. Human Albumin ELISA Quantitation kit (Bethyl Laboratory, Montgomery, TX, USA).

### 2.14. Cytochrome P450 Activity Assay

1. Cytochrome P450 2B fluorescent detection kit (Sigma-Aldrich).
2. Phenobarbital sodium (Sigma-Aldrich).
3. Ultrasonic crusher.

## 3. Methods

### 3.1. Generation of iPS Cells

Adenovirus vector plasmids for mCAT1 were transfected into HEK293 cells. The mCAT1 adenoviruses were isolated from these cells by three freeze–thaw cycles, purified using the adenovirus purification kit, and stored at –80°C. The titer of the vector stocks was determined using the Adeno-X rapid titer kit.

The replication-incompetent MMLV-based retrovirus vector pMXs was used for the ectopic expression of human OCT3/4, SOX2, c-Myc, and KLF4. Recombinant retroviruses were generated by transfection of vectors into the Plat-E packaging system, followed by incubation in FBM medium supplemented with FGM-2 SingleQuots. Between 48 and 72 h after transfection, the supernatant of an Plat-E culture was collected several times at intervals over a period of at least 4 h and passed through a 0.45-μm filter.

Fibroblasts were obtained from human neonatal foreskin, by biopsy under informed consent followed by culture in FBM supplemented with FGM-2 SingleQuots. Three days before the four-gene introduction, cells were seeded at $10^3$–$10^4$ cells/cm$^2$ into 10-cm cell culture dishes. Ten to eighteen hours later, the cells were mixed with the mCAT1 adenovirus vector solution in 500 μl Hanks' balanced salt solution and incubated at room temperature for 30 min. The cells were then added to 2 ml of medium and cultured for 48 h. Subsequently, the cells were incubated in 2 ml of retrovirus/Polybrene solution at 37°C for 4 h to overnight. The medium was changed from retrovirus/Polybrene solution to MEF-CM after infection. The medium was changed every 1–2 days.

### 3.2. Maintenance of Human iPS Cells

Colonies induced by the ectopic expression of the four genes (OCT3/4, SOX2, c-Myc, and KLF4) were isolated several days, from Day 17 to Day 33 after the four-gene transduction by using a cloning cylinder or forceps. Human iPS cells were established by repeated cloning on a feeder layer of MEFs in human iPS cell medium or on Matrigel -coated plates in mTeSR1 medium.

For passaging, the cells were usually treated with the ROCK inhibitor Y-27632 to prevent apoptosis. The cells were washed with Hanks' balanced salt solution, incubated in 0.25% trypsin–EDTA at 37°C for 3 min, and then added to the culture medium. The cells were centrifuged at $300 \times g$ at 4°C and the supernatant was removed. The cells were resuspended in culture medium with 5–20 µM Y-27632. The passages were split at 1:4 to 1:6.

Human iPS cells were frozen using Cell Freezing Solution for hES cells according to the manufacturer's manual.

### 3.3. Karyotype Analysis

Karyotype analysis of long-term cultured human iPS cells was performed using multicolor FISH analysis. Human iPS cells were pretreated with 0.02 µg/ml Colcemid for 2–3 h, incubated with 0.06–0.075 M KCl for 20 min, and then fixed with Carnoy's fixative. For multicolor FISH analysis, the cells were hybridized with a multicolor FISH probe and analyzed using a DMRA2 fluorescence microscope.

### 3.4. Real-Time PCR Gene Expression Analyses

For the quantitative analysis of gene expression in human iPS cells, total RNA was extracted from colonies using the RecoverAll Total Nucleic Acid Isolation kit. cDNA was synthesized from total RNA using the SuperScriptIII First-Strand Synthesis System. After cDNA preparstion, genes of interest were amplified using TaqMan preamp. Real-time quantitative PCR was performed with an ABI Prism 7900HT using the following PCR primer sets NANOG, Hs02387400_g1; TERT, Hs00162669_m1; GDF3, Hs00220998_m1; CYP26A1, Hs00175627_m1; GAPDH, Hs99999905_m1; DNMT3B, Hs00171876_ml; FOXD3, Hs00255287_s1; ZFP42, Hs01938187_s1; and TDGF1, Hs02339499_g1. Standard curves were generated for each primer pair. All expression values were normalized to GAPDH.

### 3.5. DNA Microarray Analysis

Microarray analysis was carried out using Human Genome U133 Plus 2.0 gene expression arrays or Whole Human Genome Oligo microarrays. Total RNA was extracted from cells with RNeasy. The analyses were performed according to Affymetrix or Agilent technical protocols. Data from these experiments and the GEO database were analyzed with GeneSpring GX 10.0 software.

Data obtained from the public database GEO for the hES cell line H14 (data sets GSM151739 and GSM151741) were used as representatives of human ES cells for comparison purposes.

For the cluster analyses, data from Human Genome U133 Plus 2.0 gene expression arrays were compared with DNA microarray data from the GEO database for the human ES cell line Sheff 4 cultured on MEF (GSM194307, GSM194308, GSM194309), or on matrigel (GSM194313, GSM194314), for the human ES cell line H14 cultured on MEF (GSM151739, GSM151741), and for three fibroblast cultures (GSM96262, GSM96263 and GSM96264).

For the Pearson correlation coefficient, data from Whole Human Genome Oligo microarrays were compared with data from the GEO database for the human ES cell lines (GSM194390, GSM194391, GSM194392) and a human iPS cell line 201B7 (GSM241846) established from adult fibroblasts (GSM242095).

The gene set was defined by the International Stem Cell Initiative (1, 4).

### 3.6. Alkaline Phosphatase Staining

The cells were fixed with 10% neutral-buffered formalin solution at room temperature for 5 min, washed with PBS, and incubated with alkaline phosphatase substrate (One-Step NBT/BCIP) at room temperature for 20–30 min. Cells positive for alkaline phosphatase activity were stained blue-violet.

### 3.7. Immuno-cytochemistry

Cultured cells were fixed with 10% formaldehyde for 10 min and blocked with 0.1% gelatin/PBS at room temperature for 1 h. The cells were incubated overnight at 4°C with primary antibodies against SSEA-3 (MC-631), SSEA-4 (MC813-70), TRA-1-60 (ab16288), TRA-1-81 (ab16289), CD9 (M-L13), CD24 (ALB9), Thy1 (5E10), or Nanog (MAB1997). For NANOG staining, cells were permeabilized with 0.1% Triton X-100/PBS before blocking. The cells were then washed with PBS three times and then incubated with AlexaFluor 488-conjugated secondary antibodies and Hoechst 33258 at room temperature for 1 h. After further washing, fluorescence was detected with an Axiovert 200 M microscope.

### 3.8. Primers for RT-PCR Gene Expression Analysis

The PCR primers listed in Table 1 were used for the RT-PCR analysis of HPRT1, NANOG, TERT, SALL4, ZFP42, GDF3, DNMT3B, TDGF1, GABRB3, and CYP26A1. The PCR primers in Table 2 were used for the analysis of gene silencing in the ectopically expressed genes.

### 3.9. Teratoma Formation

The iPS cell suspension ($0.5–2 \times 10^6$ cells/mouse) was injected into the medulla of the left testis of 7- to 8-week-old SCID mice (CB17) using a Hamilton syringe. After 8–10 weeks, the teratomas were excised after perfusion with PBS followed by 10% buffered formalin and subjected to histological analysis.

### 3.10. Histology and Immuno-histochemistry

Teratomas were embedded in mounting medium, and 10 μm frozen sections were prepared. Serial sections were stained with hematoxylin–eosin (HE) to visualize the general morphology.

For the detection of cartilage, Alcian blue staining was employed alone or in combination with HE.

For immunostaining, sections were treated with Immunoblock for 30 min to block nonspecific binding. Slides were incubated with the following primary antibodies: anti-nestin polyclonal antibody (PRB-570 C, 1:300), anti-type II collagen polyclonal antibody (LB-1297, 1:200), anti-smooth muscle actin polyclonal antibody (RB-9010-R7, 1:1), anti-$\alpha$-fetoprotein polyclonal antibody (A0008, 1:500), anti-MUC-1 polyclonal antibody (RB-9222-P0, 1:100), and anti-human nuclei monoclonal antibody (HuNu) (MAB1281, 1:300). For type II collagen, the sections were incubated with hyaluronidase (25 mg/ml) for 30 min before treatment with primary antibodies. Antigen locations were visualized using appropriate secondary antibodies (Alexa Fluor 594 and 688, 1:600). Nuclei were stained with DAPI (InnoGenex). Immunostained teratoma sections were analyzed by fluorescence microscopy (Axio Imager Z1).

### 3.11. Methylation Analysis

The promoter regions of NANOG and OCT3/4 were analyzed for methylation of individual CpG sites. Ten nanograms of bisulfite-treated genomic DNA was PCR-amplified with primers containing a T7 promoter (Table 3), and transcripts were treated with RNase A. Methylation of individual CpG sites was assessed using a MALDI-TOF mass spectrometry-based method (Epityper). Mass spectra were collected using a MassARRAY mass spectrometer. Spectra were analyzed using proprietary peak picking and signal-to-noise calculations.

### 3.12. Southern Blot Analysis and Genomic PCR

To prepare ($^{32}$P)-labeled probes, cDNA fragments of OCT3/4, SOX2, and KLF4 were enzymatically extracted from the corresponding pMXs vector plasmids (*Xho*I for OCT3/4, *Not*I for Sox2, and *Pst*I for KLF4). These crude fragments were purified by 0.8% agarose gel electrophoresis with the gel extraction kit followed by ($^{32}$P)-labeling. Genomic DNAs were prepared from the human iPS cells and its parental fibroblasts. Five microgram of each genomic DNA was digested with *Kpn*I. The fragments were separated on a 0.8% agarose gel, blotted onto HybondXL membrane, and hybridized with ($^{32}$P)-labeled probes. Genomic DNAs prepared from the human iPS cells and its parental fibroblasts were subjected to PCR to detect the c-Myc transgene using a primer set designed to amplify the c-Myc gene, including its second intron.

### 3.13. SNP Genotyping

SNP genotyping was performed using the GeneChip Human Mapping 500 K Array Set according to the manufacturer's protocol. Two aliquots of 250 ng of DNA for each sample were digested with *Nsp*I and *Sty*I. Each enzyme preparation was hybridized to the corresponding SNP array (262,000 and 238,000 on the *Nsp*I and *Sty*I array respectively). The 93% call rate threshold at $P = 0.33$ with the Dynamic Model algorithm138 was used in individual assays.

### 3.14. HLA Typing

The detection of alleles at the human leukocyte antigen HLA-A, -B, -Cw, -DRB1, DQB1, and DPB1 loci was carried out by amplifying the target genes by PCR using biotin labeled primers. Amplified fragments were denatured and hybridized with sequence-specific oligonucleotide probes conjugated to color-coded microbeads. Hybridization of amplified DNA was identified by cytometry dual-laser analysis.

### 3.15. Hepatic Differentiation of Human iPS Cells

For endoderm induction, iPS cells were incubated for 24 h in RPMI 1640 medium, supplemented with 0.5 mg/ml albumin fraction V and 100 ng/ml activin A. On the following 2 days, 0.1 and 1% insulin-transferrin-selenium was added to this medium. Following activin A treatment, the differentiated human iPS cells were cultured in Hepatocyte Culture Medium (HCM) containing 30 ng/ml FGF4 and 20 ng/ml BMP2 for 4 days. Then, the differentiated cells were incubated in HCM containing 20 ng/ml HGF and 20 ng/ml KGF for 6 days, in HCM containing 10 ng/ml oncostatin M plus 0.1 μM dexamethasone for 5 days, and in DMEM containing N2, B27, 1 mM/l GlutaMAX™, 1% nonessential amino acids, and 0.1 mM β-mercaptoethanol for another 3 days.

### 3.16. Immuno-fluorescence Assay

The cells were fixed with 4% paraformaldehyde for 15 min, blocked and permeabilized with 10% normal goat or rabbit serum and 0.2% triton X-100 in PBS at room temperature for 45 min. Then, the cells were incubated with primary antibody overnight at 4°C. For surface marker staining, the permeabilization step was omitted. The cells were washed with PBS between each step. The following primary antibodies were diluted at 1:200: human rabbit anti OCT3/4, goat anti NANOG, mouse anti SSEA4, mouse anti TRA-1-60, mouse anti TRA-1-81, rabbit anti Ki67, goat anti SOX17, rabbit anti FOXA2, mouse anti CK18, and mouse anti AFP. The rabbit anti ALB antibody was diluted at 1:500. The antibodies against human rabbit anti AAT and rabbit anti CYP3A4 were diluted at 1:200. After five washes with PBS, FITC-conjugated or TRITC-conjugated secondary antibody was diluted at 1:200 and applied to the cells overnight at 4°C. Then, 1 μg/ml 4,6-diamidino-2-phenylindole was used to stain the cell nuclei. The corresponding isotype antibody or the normal serum from the same species with the primary antibody was used as a negative control.

### 3.17. PAS Stain for Glycogen

The cells in culture dishes were fixed in 4% paraformaldehyde and were treated with 5 g/l amylase for 15 min at 37°C. The further assay was performed according to the manufacturer's instructions.

### 3.18. Urea Nitrogen Kinetic Quantitative Determination

The cells were trypsinized and counted with a hemocytometer. The sample supernatants were stored at –20°C and the assay was performed according to the manufacturer's instructions. Absorbance was measured using a Synergy HT Multi-Detection

Microplate Reader with Gen5 software. Urea production was normalized to the total cell numbers.

**3.19. Albumin Secretion ELISA Assay**

The human albumin content in the supernatant was determined by the Human Albumin ELISA Quantitation kit according to the manufacturer's instructions. The cells were trypsinized and counted with a hemocytometer. The albumin secretion was normalized to the total cell numbers.

**3.20. Cytochrome P450 Activity Assay**

For the cytochrome P450 activity assay, 200 µg/ml phenobarbital sodium was added to the differentiated human iPS cells during the last 3 days and to human hepatocytes for 3 days. The medium was refreshed everyday. The samples were homogenized with an Ultrasonic crusher and the assay was performed according to the manufacturer's instructions. The fluorescence was measured with the BioTek Multi-Detection Microplate Reader, using Gen5 software. The maximum kinetic reaction velocity was normalized to the total cell number.

## Acknowledgment

This work was supported in part by a Grant-in-Aid for the Third-Term Comprehensive 10-Year Strategy for Cancer Control (T.O.); Health Science Research Grants for Research on the Human Genome and Regenerative Medicine from the Ministry of Health, Labor, and Welfare of Japan (T.O.); and a Grant from the Program for Promotion of Fundamental Studies in Health Sciences of the National Institute of Biomedical Innovation (NiBio) (T.I.).

## References

1. Masaki H, Ishikawa T, Takahashi S, Okumura M, Sakai N, Haga M, Kominami K, Migita H, McDonald F, Shimada F, Sakurada K. (2007) Heterogeneity of pluripotent marker gene expression in colonies generated in human iPS cell induction culture. Stem Cell Res. 1, 105–115.

2. Song Z, Cai J, Liu Y, Zhao D, Yong J, Duo S, Song X, Guo Y, Zhao Y, Qin H, Yin X, Wu C, Che J, Lu S, Ding M, Deng H. (2009) Efficient generation of hepatocyte-like cells from human induced pluripotent stem cells. Cell Res. 19, 1233–1242.

3. Kitamura T., Koshino Y., Shibata F., Oki T., Nakajima H., Nosaka T., Kumagai H. (2003) Retrovirus-mediated gene transfer and expression cloning: Powerful tools in functional genomics. Exp. Hematol. 31, 1007–1014.

4. Dvorak P, Emanuelsson K, Fleck RA, Ford A, Gertow K, Gertsenstein M, Gokhale PJ, Hamilton RS, Hampl A, Healy LE, Hovatta O, Hyllner J, Imreh MP, Itskovitz-Eldor J, Jackson J, Johnson JL, Jones M, Kee K, King BL, Knowles BB, Lako M, Lebrin F, Mallon BS, Manning D, Mayshar Y, McKay RD, Michalska AE, Mikkola M, Mileikovsky M, Minger SL, Moore HD, Mummery CL, Nagy A, Nakatsuji N, O'Brien CM, Oh SK, Olsson C, Otonkoski T, Park KY, Passier R, Patel H, Patel M, Pedersen R, Pera MF, Piekarczyk MS, Pera RA, Reubinoff BE, Robins AJ, Rossant J, Rugg-Gunn P, Schulz TC, Semb H, Sherrer ES, Siemen H, Stacey GN, Stojkovic M, Suemori H, Szatkiewicz J, Turetsky T, Tuuri T, van den Brink S, Vintersten K, Vuoristo S, Ward D, Weaver TA, Young LA, Zhang W. (2007) Characterization of human embryonic stem cell lines by the International Stem Cell Initiative. Nat Biotechnol. 25, 803–816.

# Chapter 10

# Efficient Hepatic Differentiation from Human iPS Cells by Gene Transfer

## Kenji Kawabata, Mitsuru Inamura, and Hiroyuki Mizuguchi

## Abstract

Establishment of protocols for the differentiation of hepatic cells from human embryonic stem (ES) and induced pluripotent stem (iPS) cells could contribute to regenerative cell therapies or drug discovery and development. However, the differentiation efficiency of endoderm-derived cells, such as hepatic cells, from human ES and iPS cells is poor because hepatic cells are differentiated via multiple lineages including endodermal cells, hepatic progenitor cells, and mature hepatocytes. We show here the protocols for efficient hepatic differentiation from human ES and iPS cells by adenovirus vector-mediated gene transfer.

**Key words:** ES cells, iPS cells, Hepatocytes, Adenovirus vector, Regenerative medicine, Drug development

## 1. Introduction

In vertebrate development, the liver is derived from the primitive gut tube, which is formed by a flat sheet of cells called the definitive endoderm (1, 2). Afterward, the definitive endoderm is separated into the liver buds and differentiated into hepatoblasts. The hepatoblasts can differentiate into both mature hepatocytes and cholangiocytes. Each step of cell growth and differentiation is tightly regulated by intra- and extracellular signaling (3). Activin A, fibroblast growth factors (FGFs), bone morphogenic protein (BMP), hepatocyte growth factor (HGF), and oncostatin M (OSM) are the most essential extracellular signaling molecules. At the intracellular level, the liver-enriched transcription factors, i.e., hepatocyte nuclear factors (HNFs), CCAAT enhancer binding protein (C/EBP) $\alpha$ and $\beta$, and hematopoietically expressed homeobox (HEX), are required for the hepatic differentiation (4, 5). Among these

Takahiro Ochiya (ed.), *Liver Stem Cells: Methods and Protocols*, Methods in Molecular Biology, vol. 826,
DOI 10.1007/978-1-61779-468-1_10, © Springer Science+Business Media, LLC 2012

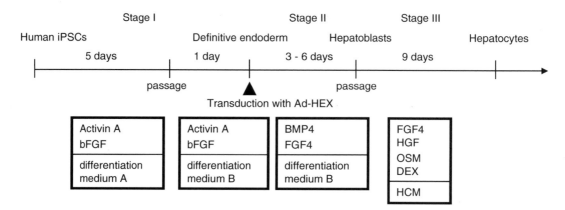

Fig. 1. A strategy for the differentiation of human iPS cells into hepatoblasts and hepatocytes. A schematic representation illustrating the procedure for differentiation of human iPS cells into hepatocytes is shown.

transcription factors, Hex is known to function at the earliest stage in hepatic differentiation (6). Targeted deletion of the HEX gene in the mouse results in embryonic lethality and a loss of the fetal liver parenchyma (7, 8). The hepatic genes, such as albumin, HNF4a, and prospero-related homeobox 1 (PROX1), are transiently expressed in the definitive endoderm of HEX-null embryos, and further morphogenesis of the hepatoblasts does not occur (9). Together, these findings underscore that HEX is essential for the definitive endoderm to adopt a hepatic cell fate.

Here, we show the protocol for the efficient differentiation of hepatoblasts from human ES and iPS cells. Our strategy is based on an imitation of in vivo liver development (Fig. 1). We have found that differentiation of hepatoblasts from the human ES and iPS cell-derived definitive endoderms, but not from undifferentiated human ES and iPS cells, could be facilitated by adenovirus (Ad) vector-mediated transient transduction of a HEX gene (10). Hepatoblasts derived from human iPS cells by HEX transduction were able to differentiate into functional hepatocytes in vitro. Furthermore, all the procedures for culture and differentiation were performed under serum/feeder cell-free chemically defined conditions. Our protocol based on Ad vector-mediated transient transduction under chemically defined conditions would provide a platform for drug screening as well as safe regenerative cell therapies.

## 2. Materials

### 2.1. Adenovirus Vectors

1. The human HEX cDNA (GenBank Accession No. BC014336) (Invitrogen, Carlsbad, CA).
2. Shuttle plasmid pHMEF5 (11).
3. Vector plasmid pAdHM41-K7 (12).

**2.2. Cells**

1. Human iPS cells (see Note 1).

2. Mitomycin C-inactivated mouse embryonic fibroblasts (MEF) (Hygro-Resistant Strain C57/BL6) (Millipore, Bedford, MA) (see Note 1).

3. HepG2 cells.

**2.3. Medium and Growth Factors**

1. Defined serum-free medium (hESF9): hESF-GRO medium (Cell Science & Technology Institute, Sendai, Japan) supplemented with 10 μg/ml human recombinant insulin, 5 μg/ml human apotransferrin, 10 μM 2-mercaptoethanol, 10 μM ethanolamine, 10 μM sodium selenite, oleic acid conjugated with fatty acid-free bovine albumin, 10 ng/ml bFGF, and 100 ng/ml heparin (all from Sigma, St. Louis, MO).

2. Laminin from the Engelbreth-Holm-Swarm murine sarcoma basement membrane (Sigma).

3. Twelve-well culture plate (Sumitomo Bakelite, Tokyo, Japan).

4. Laminin-coated tissue culture 12-well plate: Dilute laminin in PBS for a final dilution of 1:50. Add 1 ml of laminin solution to coat each well of a 12-well plate. Incubate the plates for 3–24 h at 37°C. Remove laminin solution and wash the well with PBS immediately before use.

5. Accutase (Invitrogen).

6. Differentiation medium A: hESF-GRO medium (Cell Science & Technology Institute) supplemented with 10 μg/ml human recombinant insulin, 5 μg/ml human apotransferrin, 10 μM 2-mercaptoethanol, 10 μM ethanolamine, 10 μM sodium selenite, and 0.5 mg/ml fatty acid-free bovine albumin (BSA) (Sigma).

7. bFGF (Sigma).

8. Activin A (R&D Systems, Minneapolis, MN).

9. Trypsin–EDTA: 0.0125% trypsin, 0.01325 mM EDTA (Invitrogen).

10. Trypsin inhibitor A: Differentiation medium A supplemented with 0.1% soybean trypsin inhibitor (Sigma).

11. Differentiation medium B: hESF-DIF (Cell Science & Technology Institute) medium supplemented with 10 μg/ml human recombinant insulin, 5 μg/ml human apotransferrin, 10 μM 2-mercaptoethanol, 10 μM ethanolamine, 10 μM sodium selenite, and 0.5 mg/ml fatty acid-free BSA.

12. FGF4 (R&D Systems).

13. BMP4 (R&D Systems).

14. Trypsin inhibitor B: Differentiation medium B supplemented with 0.1% soybean trypsin inhibitor (Sigma).

15. Hepatocyte culture medium (HCM) supplemented with SingleQuots (Lonza, Walkersville, MD).

16. HGF (R&D Systems).

**Table 1**
**List of Taqman gene expression assays**

| Gene | Assay ID |
| --- | --- |
| AFP | Hs01040607_m1 |
| ALB | Hs00910225_m1 |
| CYP3A4 | Hs00430021_m1 |
| CYP7A1 | Hs00167982_m1 |
| CYP2D6 | Hs02576168_g1 |

17. Oncostatin M (OSM) (R&D Systems).

18. Dexamethasone (Sigma).

19. Type I collagen (Nitta Gelatin, Osaka, Japan).

20. Type I collagen-coated 12-well plate ($15 \ \mu g/cm^2$): Dilute type I collagen in PBS for a final dilution of 1:50. Add 1 ml of type I collagen solution to coat each well of a 12-well plate. Incubate the plates for 3–24 h at 37°C. Remove type I collagen solution immediately before use.

*2.4. Analysis*

1. Human fetal (22–40 weeks old) liver total RNA (Clontech Laboratories, Mountain View, CA).

2. Human adult (51 years old) liver total RNA (Clontech Laboratories).

3. RNeasy Plus Mini kit (Qiagen, Hilden, Germany).

4. Superscript VILO cDNA synthesis kit (Invitrogen).

5. Taqman gene expression assays (Applied Biosystems, Foster City, CA): The primer sequences are described in Table 1.

6. ABI PRISM 7700 Sequence Detector (Applied Biosystems).

7. P450-GloTM CYP3A4 Assay Kit (Promega, Madison, WI).

8. Rifampicin (Sigma).

9. Dimethyl sulfoxide (Sigma).

10. Luminometer (Berthold, Tokyo, Japan).

# 3. Methods

*3.1. Adenovirus Vector Construction*

1. Ad vectors were constructed by an improved in vitro ligation according to the method of Mizuguchi and Kay. (13, 14). The human HEX cDNA was inserted into pHMEF5, which contains the human elongation factor-1α (EF-1α) promoter, resulting in pHMEF-HEX.

2. The pHMEF-HEX was digested with I-CeuI/PI-SceI and ligated into I-CeuI/PI-SceI-digested pAdHM41-K7, resulting in pAd-HEX.

3. Ad-HEX, which contains the EF-1α promoter and a stretch of lysine residues (K7) peptides in the C-terminal region of the fiber knob, was generated and purified.

4. The vector particle (VP) titer was determined by using a spectrophotometric method (15).

**3.2. In Vitro Definitive Endoderm Differentiation**

1. Prepare human iPS cells, which were maintained on MEF on a gelatin-coated 25 $cm^2$ flask in human iPS cell culture medium (see Note 1).

2. Before the initiation of cellular differentiation, change the medium of human iPS cells for the defined serum-free medium hESF9.

3. Incubate the cells in a humidified atmosphere of 10% $CO_2$ and 90% air at 37°C overnight (see Note 2).

4. For induction of definitive endoderm, remove the hESF9 medium, add 1.0 ml Accutase per 25-$cm^2$ flask, incubate for 3 min at 37°C, and remove the Accutase (see Note 3).

5. Add 10 ml of cold hESF9 medium, resuspend the human iPS cells into a single cell suspension by pipetting, and centrifuge at $267 \times g$ for 3 min at 4°C (see Note 4).

6. Aspirate the supernatant and resuspend the cells with 10 ml of cold differentiation medium A and centrifuge them at $267 \times g$ for 3 min at 4°C.

7. Repeat step 6.

8. Aspirate the supernatant, and replace the medium with warm fresh differentiation medium A supplemented with 10 ng/ml bFGF and 50 ng/ml Activin A.

9. Transfer to a laminin-coated 12-well plate in a humidified atmosphere of 10% $CO_2$ and 90% air at 37°C ($2.5 \times 10^5$ cells/well). The final volume of medium should be 1.0 ml per well (see Note 5).

10. Change the differentiation medium A supplemented with 10 ng/ml bFGF and 50 ng/ml Activin A every day.

**3.3. In Vitro Hepatoblast Differentiation**

1. After 5 days of culture, remove the medium, add 200 μl trypsin–EDTA per well, incubate the cells for 3 min at 37°C, and remove the trypsin–EDTA (see Note 6).

2. Resuspend the cell populations in 10 ml of cold trypsin inhibitor A and centrifuge them at $267 \times g$ for 3 min at 4°C.

3. Aspirate the supernatant, resuspend the cells in 10 ml of cold differentiation medium B, and centrifuge at $267 \times g$ for 3 min at 4°C.

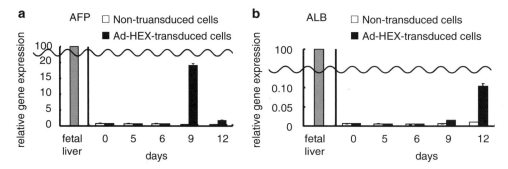

Fig. 2. Efficient hepatoblast differentiation from the human iPS cell-derived definitive endoderms by transduction of the HEX gene. Real-time RT-PCR analysis of the level of AFP (**a**) and ALB (**b**) expression in nontransduced cells and Ad-HEX-transduced cells, both of which were induced from the human iPS cell-derived definitive endoderms (day 0, 5, 6, 9, and 12). The cells were transduced with Ad-HEX at day 6 as described in Fig. 1. The data at day 6 were obtained before the transduction with Ad-HEX. The *graphs* represent the relative gene expression levels when the level in the fetal liver was taken as 100.

4. Aspirate the supernatant and replace with warm fresh differentiation medium B supplemented with 10 ng/ml bFGF and 50 ng/ml Activin A.

5. Transfer the cells to a laminin-coated tissue culture 12-well plate ($5.0 \times 10^5$ cells/well) and culture them in a humidified atmosphere of 10% $CO_2$ and 90% air at 37°C. The final volume of medium should be 1.0 ml per well (see Note 5).

6. After 24 h of culture, remove the medium, and add warm fresh differentiation medium B supplemented with Ad-HEX (3,000 VP/cell), 10 ng/ml FGF4, and 10 ng/ml BMP4 (R&D Systems) (see Note 7). The final volume of medium should be 500 µl per well.

7. Incubate the cells in a humidified atmosphere of 10% $CO_2$ and 90% air at 37°C for 1.5 h.

8. Remove the medium and replace with warm fresh differentiation medium B supplemented with 10 ng/ml FGF4 and 10 ng/ml BMP4, and incubate the cells in a humidified atmosphere of 10% $CO_2$ and 90% air at 37°C.

9. Change the medium every day (see Note 8).

10. After 3 and 6 days of culture in differentiation medium B, analyze the cells by RT-PCR (see Note 9) (Fig. 2).

### 3.4. In Vitro Hepatic Maturation

1. After 3 days of culture in differentiation medium B, add 200 µl trypsin–EDTA in each well, incubate the cells for 3 min at 37°C, and remove the trypsin–EDTA.

2. Resuspend the cell populations in 10 ml of cold trypsin inhibitor B and centrifuge them at $267 \times g$ for 3 min at 4°C (see Note 10).

3. Aspirate the supernatant, resuspend the cells in 10 ml of cold HCM and centrifuge them at $267 \times g$ for 3 min at 4°C.

4. Aspirate the supernatant, and replace with warm fresh HCM supplemented with SingleQuots, 10 ng/ml FGF-4, 10 ng/ml HGF, 10 ng/ml, and $10^{-7}$ M dexamethasone.

5. Transfer into two wells of a type I collagen-coated tissue culture 12-well plate and incubate the cells in a humidified atmosphere of 10% $CO_2$ and 90% air at 37°C. The final volume of medium should be 1.0 ml per well (see Note 5).

6. Change the medium every 2 days.

7. After 9 days of culture in HCM, analyze the cells by RT-PCR and measure the cytochrome P450 activity (see Notes 9 and 11) (Fig. 3).

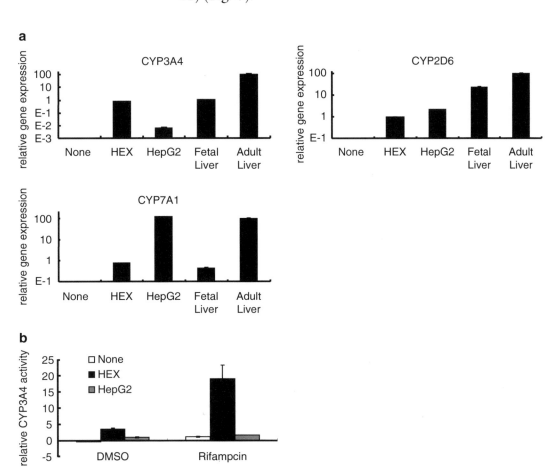

Fig. 3. Cytochrome P450 isozymes in human iPS cell-derived hepatocytes. (**a**) Real-time RT-PCR analysis of CYP3A4, CYP7A1, and CYP2D6 expression in human iPS cell-derived nontransduced cells (day 18), Ad-HEX-transduced cells (day 18), and fetal and adult liver tissues. (**b**) Induction of CYP3A4 by rifampicin in human iPS cell-derived nontransduced cells, Ad-HEX-transduced cells, and the HepG2 cell line. Data are presented as the mean ± SD from triplicate experiments. The *graphs* represent the relative gene expression level when the level in the adult liver is taken as 100. Abbreviations: *NONE* nontransduced cells, *LacZ* Ad-LacZ-transduced cells, *HEX* Ad-HEX-transduced cells, *DMSO* dimethyl sulfoxide.

## 4. Notes

1. Culture human iPS cells to maintain the undifferentiated states according to the original protocol (16, 17). Basically, human ES cells can be cultured, handled, and differentiated using the same protocol as human iPS cells described here.

2. Proceed to step 3 within 48 h. Attachment efficiency will be reduced if passage is performed after more than 48 h of culture in hESF9 medium.

3. By this operation, the feeder cells are removed and only the human iPS cells remain in the flask.

4. Determine the number of cells using a hemocytometer and adjust the concentration precisely. An excessive number of cells per well results in the presence of undifferentiated cells after 5 days of culture with differentiation medium. Also, strain the cell suspension with a cell strainer to obtain a uniform single cell suspension, since cell clusters will result in the appearance of undifferentiated cells after 5 days of culture with differentiation medium.

5. Be sure to gently shake the plate left to right and back to front to obtain evenly distributed cells.

6. A low concentration of trypsin–EDTA can reduce cell damage by passage and promote cell survival. Detach the cells by brushing the medium on the cells.

7. Vortex the 1.5-ml tube supplemented with Ad-HEX.

8. Proceed to Subheading 3.4. for induction of hepatocytes after 3 days of culture in differentiation medium B.

9. Total RNA was isolated from human iPS cells, their derivatives, and HepG2 cells using an RNeasy Plus Mini kit. cDNA was synthesized using 500 ng of total RNA with a Superscript VILO cDNA synthesis kit. Real-time PCR was performed with Taqman gene expression assays using an ABI PRISM 7700 Sequence Detector. Relative quantification was performed against a standard curve, and the values were normalized against the input determined for the housekeeping gene, glyceraldehyde-3-phosphate dehydrogenase (GAPDH). The primer sequences used in these methods are described in Table 1.

10. Do not dissociate the cell clusters into single cells. Passage the cells as the cell clumps.

11. To measure cytochrome P450 3A4 activity, lytic assays was performed by using a P450-GloTM CYP3A4 Assay Kit. For the cytochrome P450 3A4 activity assay, Ad-HEX-transduced cells and nontransduced cells as well as HepG2 cells were

treated with rifampicin, which is the substrate for CYP3A4, at a final concentration of 25 µM or DMSO for 72 h. The fluorescence was measured with a luminometer according to the manufacturer's instructions. HepG2 cells were cultured as per the instructions.

## Acknowledgment

This study was supported by grants from the Ministry of Education, Sports, Science and Technology of Japan and by grants from the Ministry of Health, Labor, and Welfare of Japan.

## References

1. Lavon, N. and Benvenisty, N. (2005) Study of hepatocyte differentiation using embryonic stem cells. *J Cell Biochem* 96, 1193–1202.

2. McLin, V.A. and Zorn, A.M. (2006) Molecular control of liver development. *Clin Liver Dis* 10, 1–25.

3. Snykers, S., De Kock, J., Rogiers, V., and Vanhaecke, T. (2009) In vitro differentiation of embryonic and adult stem cells into hepatocytes: State of the art. *Stem Cells* 27, 577–605.

4. Kyrmizi, I., Hatzis, P., Katrakili, N., Tronche, F., Gonzalez, F.J., and Talianidis, I. (2006) Plasticity and expanding complexity of the hepatic transcription factor network during liver development. *Genes Dev* 20, 2293–2305.

5. Hunter, M.P., Wilson, C.M., Jiang, X., Cong, R., Vasavada, H., Kaestner, K.H., and Bogue, C.W. (2007) The homeobox gene Hhex is essential for proper hepatoblast differentiation and bile duct morphogenesis. *Dev Biol* 308, 355–367.

6. Bogue, C.W., Ganea, G.R., Sturm, E., Ianucci, R., and Jacobs, H.C. (2000) Hex expression suggests a role in the development and function of organs derived from foregut endoderm. *Dev Dyn* 219, 84–89.

7. Martinez Barbera, J.P., Clements, M., Thomas, P., Rodriguez, T., Meloy, D., Kioussis, D., and Beddington, R.S. (2000) The homeobox gene Hex is required in definitive endodermal tissues for normal forebrain, liver and thyroid formation. *Development* 127, 2433–2445.

8. Keng, V.W., Yagi, H., Ikawa, M., Nagano, T., Myint, Z., Yamada, K., Tanaka, T., Sato, A., Muramatsu, I., Okabe, M., Sato, M., and Noguchi, T. (2000) Homeobox gene Hex is essential for onset of mouse embryonic liver development and differentiation of the monocyte lineage. *Biochem Biophys Res Commun* 276, 1155–1161.

9. Bort, R., Signore, M., Tremblay, K., Martinez Barbera, J.P., and Zaret, K.S. (2006) Hex homeobox gene controls the transition of the endoderm to a pseudostratified, cell emergent epithelium for liver bud development. *Dev Biol* 290, 44–56.

10. Inamura, M., Kawabata, K., Takayama, K., Tashiro, K., Sakurai, F., Katayama K., Toyoda, M., Akutsu, H., Miyagawa, Y., Okita, H., Kiyokawa, N., Umezawa, A., Hayakawa, T., Kusuda-Furue, M., and Mizuguchi, H. Efficient generation of hepatoblasts from human ES cells and iPS cells by transient overexpression of homeobox gene HEX. in press.

11. Kawabata, K., Sakurai, F., Yamaguchi, T., Hayakawa, T., and Mizuguchi, H. (2005) Efficient gene transfer into mouse embryonic stem cells with adenovirus vectors. *Mol Ther* 12, 547–554.

12. Koizumi N, Mizuguchi H, Utoguchi N, Watanabe Y, and Hayakawa T. (2003) Generation of fiber-modified adenovirus vectors containing heterologous peptides in both the HI loop and C terminus of the fiber knob. *J Gene Med* 5, 267–276

13. Mizuguchi H and Kay M.A. (1998) Efficient construction of a recombinant adenovirus vector by an improved in vitro ligation method. *Hum Gene Ther* 9, 2577–2583.

14. Mizuguchi H and Kay M.A. (1999) A simple method for constructing E1- and E1/

E4-deleted recombinant adenoviral vectors. *Hum Gene Ther* 10, 2013–2017.

15. Maizel JV, Jr., White DO, and Scharff M.D. (1968) The polypeptides of adenovirus. I. Evidence for multiple protein components in the virion and a comparison of types 2, 7A, and 12. *Virology* 36, 115–125.

16. Takahashi K, Tanabe K, Ohnuki M, Narita M, Ichisaka T, Tomoda K, and Yamanaka S. (2007) Induction of pluripotent stem cells from adult human fibroblasts by defined factors. *Cell* 131, 861–872.

17. Nagata S, Toyoda M, Yamaguchi S, Hirano K, Makino H, Nishino K, Miyagawa Y, Okita H, Kiyokawa N, Nakagawa M, Yamanaka S, Akutsu H, Umezawa A, and Tada T. (2009) Efficient Reprogramming of Human and Mouse Primary Extra-Embryonic Cells to Pluripotent Stem Cells. *Genes Cells* 14, 1395–404.

# Chapter 11

# "Tet-On" System Toward Hepatic Differentiation of Human Mesenchymal Stem Cells by Hepatocyte Nuclear Factor

## Goshi Shiota and Yoko Yoshida

## Abstract

"Tet-On" system requires two DNA constructs: the first one is a transcription regulatory unit, rtTA and the second construct is the responsive element *Escherichia coli* sequences (*tetO*) linked to Pcmv driven target gene. In the absence of inducing agent doxycycline (Dox), a tetracycline derivative, rtTA does not bind to or binds weakly to operator sequences of *tetO*; therefore, no target gene is transcribed. However, in the presence of Dox, tTA binds to *tetO* and pcmv, which in turn activates the target gene. In general, the induction of transgene by Dox is rapid and can occur within hours in some systems, offering an advantage over the original tTA system for studying acute effects of transgenes. Recently, we have established a Tet-regulated expression system for hepatocyte nuclear factor 3β (HNF3β) to investigate the potency of hepatic differentiation of human mesenchymal stem cells (MSC) by HNF3β.

**Key words:** Tetracycline-inducible system, HNF3β, Mesenchymal stem cell, Hepatic differentiation, Tet repressor, Tetracycline operator 2

## 1. Introduction

Several inducible systems are currently available. The most widely used externally regulatable transgenic system is based on the tetracycline-controlled transcription (1–4). There are basic variants; one is the tetracycline-controlled transactivator (rTA) system ("Tet-Off" system) and the other reverse tetracycline-controlled transcriptional activator (rtTA) system ("Tet-On" system). "Tet-On" system, which was developed by Gossen and coworkers (1, 2), requires two DNA constructs; the first one is a transcription regulatory unit, which is a mutant Tet repressor fused to VP16 to form rtTA. The second construct is the responsive element *tetO* sequences linked to Pcmv driven target gene, the same as in the

Takahiro Ochiya (ed.), *Liver Stem Cells: Methods and Protocols*, Methods in Molecular Biology, vol. 826,
DOI 10.1007/978-1-61779-468-1_11, © Springer Science+Business Media, LLC 2012

tTA system. However, this system works through an opposite mechanism. In the absence of inducing agent doxycycline (Dox), a tetracycline derivative, rtTA does not bind to or binds weakly to operator sequences of *Escherichia coli* sequences (*tetO*), therefore, no target gene is transcribed. However, in the presence of Dox, tTA binds to *tetO* and pcmv, which in turn activates the target gene. In general, the induction of transgene by Dox is rapid and can occur within hours in some systems, offering an advantage over the original tTA system for studying acute effects of transgenes. The schema of "Tet-on" system was shown in Fig. 1. Recently, we have established a Tet-regulated expression system for hepatocyte nuclear factor 3β (HNF3β) to investigate the potency of hepatic differentiation of human mesenchymal stem cells (MSC) by HNF3β (5).

## 2. Materials

Prepare all solutions using ultrapure water (prepared by purifying deionized water) and analytical grade reagents. Prepare and store all reagents at room temperature (unless indicated otherwise). Diligently follow all waste disposal regulations when disposing waste materials.

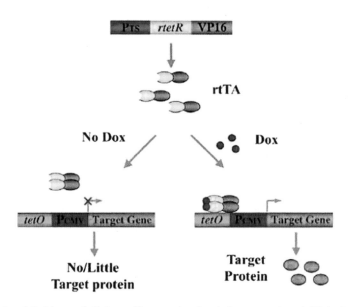

Fig. 1. A "Tet-On" system (cited from ref. 3). A specific promoter directs the expression of rtTA in the tissue or cell type of interest. In the absence of Dox, rtTA does not bind to *tetO* so there is no transcription activation of the target gene. By contrast, in the present of Dox, rtTA binds to *tetO* and initiates the transcription of the target gene.

**2.1. Preparation of pcDNA4/TO-HNF3β**

The Tet-On expression system of HNF3β was made using T-Rex™ System (Invitrogen, USA). The T-Rex™ System was stored at –20°C. Total RNA was prepared from HuH7 hepatocellular carcinoma cells using TRIzol Reagent (Invitrogen). cDNA was reverse-transcribed using SuperScript II RNase H⁻ Reverse Transcriptase (Invitrogen). The human HNF3β cDNA (GenBank NM_021784) generated with reverse-transcribed polymerase chain reaction was inserted between BamH1 site and EcoR1 site in the multiple cloning sites of pcDNA4/TO. The primers used for subcloning were HNF3β-F (BamH1 site, actcgggatccaccATGCTGCTGGGAGCG-GTGAAGATG) and HNF3β-R (EcoR1 site, acttgGAATTCatccg-gggtgccagagttagc). The resulting plasmid was designated pcDNA4/TO-HNF3β.

# 3. Methods

To establish tetracycline-induced HNF3β expression system in human UE7T-13 MSC, a regulatory plasmid, pcDNA6/TR, which encodes the Tet repressor (TetR) under the control of the human CMV promoter and an inducible expression plasmid, pcDNA4/To- HNF3β, which encodes HNF3β under the control of the strong human cytomegalovirus immediate-early (CMV) promoter and two tetracycline operator 2 (*TetO2*) (Fig. 2), were stably tranfected into these cells. The UE7T-13 bone marrow-derived MSC has been developed by being immortalized with infection with a retrovirus carrying human telomerase reverse transcriptase (hTERT) and one of the early genes of the human papilloma virus, E7 (6). Although hTERT is introduced into UE7T-13 cells, it has been reported that the differentiation potential of the cells is not affected.

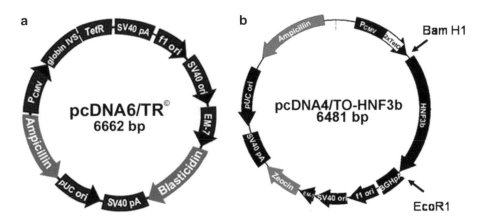

Fig. 2. Schemas of the plasmids. (**a**) pcDNA6/TR, (**b**) pcDNA4/TO-HNF3β.

**3.1. Transfection of pcDNA6/TR into EU7T-13 Cells**

1. One day before the transfection, EU7T-13 cells were split and incubated at the cell density of 25% confluency in DMEM (Nissui Pharmaceutical Co., Ltd., Japan) containing 10% FBS and penicillin–streptomycin.

2. On transfection, 10 µg linearized pcDNA6/TR was transfected into $1 \times 10^6$ cells in 500 µl DMEM without FBS or penicillin–streptomycin by means of electroporation. Electroporation was performed at 220–260 V and 1,200–1,650 µF. At 5 min after transfection, the cells were incubated in 10-cm dish and the medium was changed the next day.

3. Two days after transfection, the cells were incubated with 3 µg/ml blasticidin, which kills almost of the cells in the dish.

4. The medium containing 3 µg/ml blasticidin was changed weekly. The colonies appeared after 2–3 weeks after the start of selection and were then picked up.

5. The cloned cells were further incubated and total RNA was prepared from one portion of the cells for assessment of TetR mRNA. It is very important to select the colony expressing high levels of TetR mRNA, since in the colony with low level of TetR the expression of the gene of interest is leaky.

6. To assess TetR mRNA level in the tranfected cells, reverse transcription-polymerase chain reaction (RT-PCR) was done using GAPDH as an internal control. The forward primer was GCTTTGCTCGACCCCTTAG (1,831–1,849) and the reverse primer was TACGCGGACCCACTTTCAC (2,309–2,291). Of the ten clones, the E7-T6-10 cells express the highest level of TetR mRNA (Fig. 3), and then these cells were chosen for transfection of pcDNA/TO-HNF3β.

**3.2. Transfection of pcDNA/TO-HNF3β into the E7-T6-10 Cells**

1. Using the E7-T6-10 cells were transfected with pcDNA/TO-HNF3β in the same way with Subheading 3.1, step 1 and 2 protocols.

2. Two days after transfection, the cells were incubated with 3 µg/ml blasticidin and 100 µg/ml Zeocin™. The medium was changed weekly.

3. The colonies appeared after 2–3 weeks after the start of selection, and were then picked up.

4. The cloned cells were further incubated. The candidate colony was assessed for HNF3β mRNA expression at one day after incubation with 1 mg/ml tetracycline for 24 h. Total RNA was prepared from the cells for assessment of HNF3β mRNA.

5. To assess HNF3β mRNA level in the tranfected cells, real-time RT-PCR was done using GAPDH as an internal control. The forward primer was ACCCCAAGACCTACAGGCG (636–654) and the reverse primer was TCAGCGTCAGCATCTTGTTGG (743–723). Eleven positive clones for HNF3β were picked up (Fig. 4). Finally, the most strictly regulated cells by Dox was designated E7-H-14.

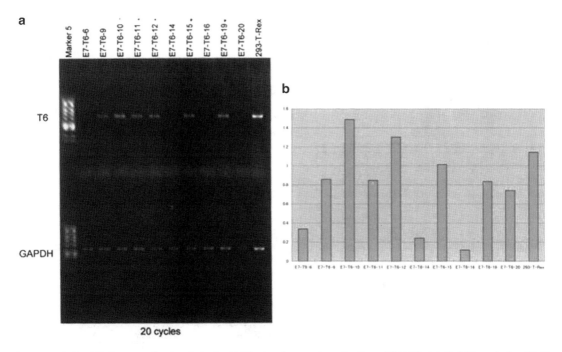

Fig. 3. Analysis of TetR expression levels in the EU7T-13 cells transfected with pcDNA6/TR. (**a**) RT-PCR analysis of TetR mRNA in the cloned cells. The cycle number of PCR was 20. (**b**) Relative expression levels of TetR mRNA in the cloned cells. The intensity of TetR mRNA is expressed as the ratio to GAPDH.

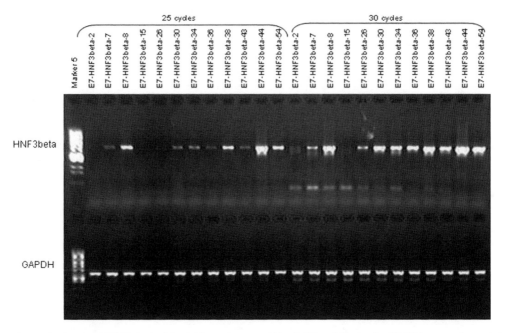

Fig. 4. Expression levels of HNF3b in the blasticidin-resistant and Zeocin™-resisitant cells. RT-PCR was performed at 15, 20, 25, and 30 cycles. The data were shown at 25 and 30 cycles.

## 4. Notes

1. To propagate and maintain the pcDNA6/TRc plasmid, the following procedures are recommended by the manufacturer

2. Resuspend the vector in 20 µl sterile water to prepare a 1 µg/µl stock solution. Store the stock solution at –20°C.

3. Use the stock solution to transform a *rec*A, *end*A *E. coli* strain such as TOP10F.

4. Select transformants on LB agar plates containing 50–100 µg/ml ampicillin or 50 µg/ml blasticidin in low salt LB.

5. Prepare a glycerol stock of each plasmid for long-term storage.

6. Preparing glycerol stock is recommended as follows by the manufacturer

7. Once you have identified the correct clone, be sure to purify the colony and make a glycerol stock for long-term storage. It is also a good idea to keep a DNA stock of your plasmid at –20°C.

8. Streak the original colony out on an LB plate containing 50 µg/ml ampicillin or 100 µg/ml blasticidin in low salt LB. Incubate the plate at 37°C overnight.

9. Isolate a single colony and inoculate into 1–2 ml of LB containing 50 µg/ml ampicillin or 50 µg/ml blasticidin in low salt LB.

10. Grow the culture to mid-log phase ($OD_{600} = 0.5$–$0.7$). Mix 0.85 ml of culture with 0.15 ml of sterile glycerol and transfer to a cryovial. Store at –80°C.

11. Preparation of tetracycline is recommended by the manufacturer as follows

12. Tetracycline (MW = 444.4) is commonly used as a broad spectrum antibiotic and acts to inhibit translation by blocking polypeptide chain elongation in bacteria. In the T-REx™ System, tetracycline is used as an inducing agent to induce transcription of the gene of interest from the inducible expression vector. Tetracycline induces transcription by binding to the Tet repressor homodimer and causing the repressor to undergo a conformational change that renders it unable to bind to the Tet operator. The association constant of tetracycline to the Tet repressor is $3 \times 10^9$ M-1 (7). Please note that the concentrations of tetracycline used to induce gene expression in the T-REx™ System are generally not high enough to be toxic to mammalian cells.

13. Determination of antibiotic sensitivity is recommended by the manufacturer

    To successfully generate a stable cell line expressing the Tet repressor and your protein of interest, you need to determine the minimum concentration of each antibiotic (blasticidin and Zeocin™) required to kill your untransfected host cell line. For each antibiotic, test a range of concentrations (see below) to ensure that you determine the minimum concentration necessary for your cell line. Use the protocol below to determine the minimal concentrations of Zeocin™ and blasticidin required to prevent growth of the parental cell line.

14. Plate or split a confluent plate so the cells will be approximately 25% confluent. For each antibiotic, prepare a set of 6–7 plates. Add the following concentrations of antibiotic to each plate in a set: For blasticidin selection, test 0, 1, 3, 5, 7.5, and 10 μg/ml blasticidin; for Zeocin™ selection, test 0, 50, 125, 250, 500, 750, and 1,000 μg/ml Zeocin™.

15. Replenish the selective media every 3–4 days and observe the percentage of surviving cells.

16. Count the number of viable cells at regular intervals to determine the appropriate concentration of antibiotic that prevents growth within 1–2 weeks after addition of the antibiotic.

17. Effects of Zeocin™ on sensitive and resistant cells are described by the manufacturer. Zeocin™'s method of killing is quite different from other antibiotics including blasticidin, neomycin, and hygromycin. Cells do not round up and detach from the plate. Sensitive cells may exhibit the following morphological changes upon exposure to Zeocin™: Vast increase in size (similar to the effects of cytomegalovirus infecting permissive cells), abnormal cell shape, presence of large empty vesicles in the cytoplasm (breakdown of the endoplasmic reticulum and Golgi apparatus, or other scaffolding proteins), breakdown of plasma and nuclear membrane (appearance of many holes in these membranes). Eventually, these "cells" will completely break down and only "strings" of protein remain.

    Zeocin™-resistant cells should continue to divide at regular intervals to form distinct colonies. There should not be any distinct morphological changes in Zeocin™-resistant cells when compared to cells not under selection with Zeocin™. For more information about Zeocin™ and its mechanism of action, please refer to the inducible expression vector manual.

## References

1. Gossen M, Bujard H (1992) Tight control of gene expression in mammalian cells by tetracycline-responsive promoters. Proc Natl Acad Sci USA 89:5547–4451

2. Gossen M, Freundlieb S, Bender G, et al (1995) Transcriptional activation by tetracyclines in mammalian cells. Science 268:1766–1769

3. Zhu Z, Zheng T, Lee CG et al (2002) Tetracycline-controlled transcriptional regulation systems: advances and application in transgenic animal modeling. Semin Cell Dev Biol 13:121–128

4. Berens C, Hillen W (2003) Gene regulation by tetracyclines: constraints of resistance regulation in bacteria shape TetR for application. Eur J Biochem 270:3109–3021

5. Ishii K, Yoshida Y, Akechi Y et al (2008) Hepatic differentiation of human bone marrow-derived mesenchymal stem cells by tetracycline-regulated hepatocyte nuclear factor 3beta. Hepatology 48:597–606

6. Mori T, Kiyono T, Imabayashi H et al (2005) Combination of hTERT and bmi-1, E6, or E7 induces prolongation of the life span of bone marrow stromal cells from an elderly donor without affecting their neurogenic potential. Mol Cell Biol 25:5183–5195

7. Takahashi M, Degenkolb J, Hillen W (1991) Determination of the equilibrium constant between Tet repressor and tetracycline at limiting $Mg2+$ concentrartion: a generally applicable method for effector-dependent high-affinity complexes. Anal Biochem 199:197–202

# Chapter 12

## SAMe and HuR in Liver Physiology

### Usefulness of Stem Cells in Hepatic Differentiation Research

**Laura Gomez-Santos, Mercedes Vazquez-Chantada, Jose Maria Mato, and Maria Luz Martinez-Chantar**

## Abstract

*S*-Adenosylmethionine, abbreviated as SAM, SAMe or AdoMet, is the principal methyl group donor in the mammalian cell and the first step metabolite of the methionine cycle, being synthesized by MAT (methionine adenosyltransferase) from methionine and ATP. About 60 years after its identification, SAMe is admitted as a key hepatic regulator whose level needs to be maintained within a specific range in order to avoid liver damage. Recently, in vitro and in vivo studies have demonstrated the regulatory role of SAMe in HGF (hepatocyte growth factor)-mediated hepatocyte proliferation through a mechanism that implicates the activation of the non-canonical LKB1/AMPK/eNOS cascade and HuR function. Regarding hepatic differentiation, cellular SAMe content varies depending on the status of the cell, being lower in immature than in adult hepatocytes. This finding suggests a SAMe regulatory effect also in this cellular process, which very recently was reported and related to HuR activity. Although in the last years this and other discoveries contributed to throw light into the tangle of regulatory mechanisms that govern this complex process, an overall understanding is still a challenge. For this purpose, the in vitro hepatic differentiation culture systems by using stem cells or fetal hepatoblasts are considered as valuable tools which, in combination with the methods used in current days to elucidate cell signaling pathways, surely will help to clear up this question.

**Key words:** *S*-Adenosylmethionine (SAMe), MAT (methionine adenosyltransferase), HuR, Hepatocyte, Liver, Hepatocyte differentiation, Hepatocyte proliferation, Stem cells

## 1. Introduction

### 1.1. Methionine Metabolism, Synthesis of SAMe

In the early 1950s, Cantoni, completing the studies carried out by Vincent du Vigneaud during 1930s, identified SAMe as the product of the reaction between methionine and ATP (adenosine triphosphate) capable of donating its methyl group to nicotinamide or creatine (1). Several years and studies after this discovery, an

Takahiro Ochiya (ed.), *Liver Stem Cells: Methods and Protocols*, Methods in Molecular Biology, vol. 826,
DOI 10.1007/978-1-61779-468-1_12, © Springer Science+Business Media, LLC 2012

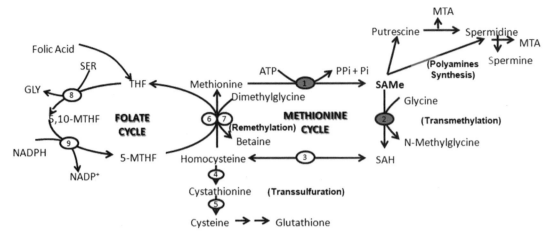

Fig. 1. Hepatic SAMe metabolism. SAMe is synthesized in the cytosol of every cell. However, SAMe synthesis and utilization occurs mainly in the liver. SAMe is generated from methionine and ATP in a reaction catalyzed by MAT (1). MTA is produced from SAMe through the polyamine biosynthetic pathway, and this compound is metabolized solely by MTA-phosphorylase to yield 5-methylthioribose-1-phosphate and adenine, a crucial step in the methionine and purine salvage pathways, respectively. In transmethylation, SAMe donates its methyl group to a large variety of acceptor molecules in reactions catalyzed by dozens of methyltransferases. The most abundant of these methyltransferases in mammalian liver is GNMT (2). SAH is generated as a product of transmethylation and is hydrolyzed to form homocysteine (Hcy) and adenosine through a reversible reaction catalyzed by SAH hydrolase (3). Hcy lies at the junction of two intersecting pathways: the *transsulfuration pathway*. In the liver, Hcy forms cysteine via a two-step enzymatic process catalyzed by cystathionineβ-synthase (CBS) (4) and cystathionase (5), which converts the sulfur atom of methionine to cysteine and glutathione; and the *remethylation pathway*: homocysteine can be remethylated to form methionine by two different enzymes, methionine synthase (MS) (6), and betaine homocysteine methyltransferase (BHMT) (7), and is coupled to the *folate cycle*. In this cycle, tetrahydrofolate (THF) is converted to 5,10-methylenetetrahydrofolate (5,10-MTHF) by the enzyme methyleneTetrahydrofolate synthase (8) and then to 5-methyl-tetrahydrofolate (5-MTHF) by the enzyme methyleneTetrahydrofolate reductase (MTHFR) (9).

integrated concept of the methionine cycle and SAMe as the link between the main metabolic pathways: polyamines synthesis, transsulfuration, transmethylation and folate metabolism, was provided (2) (Fig. 1), emerging SAMe as the major biological donor of methyl groups.

In mammals, MAT is the sole enzyme that catalyzes the formation of SAMe and the product of two different genes, *MAT1A* and *MAT2A* (3). In the adult organism, *MAT1A* is mainly expressed in the hepatocytes encoding MATI/III enzymes, whereas *MAT2A*, which codes for MATII, is expressed in all tissues. Because of differences in the regulatory and kinetic properties of the various MAT isoforms, the gene product of *MAT1A* is more efficient in the synthesis of SAMe than the gene product of *MAT2A* (4–6). In consequence, although all mammalian cells can synthesize SAMe, the liver is the principal organ for conversion of dietary methionine into SAMe and where up to 85% of all transmethylation reactions occur (7).

**1.2. SAMe in Liver Pathogenesis**

The effect the alteration of methionine metabolism and, in consequence, of cellular SAMe content, has on liver pathophysiology was discovered several decades ago. In 1932, Best's group demonstrated that rats fed with a diet deficient in methyl groups, such as methionine and choline, spontaneously develop liver steatosis (fatty liver) within a few weeks (8). In case the diet continues, rats progressively develop NASH (non-alcoholic steatohepatitis; inflamed fatty liver), fibrosis, cirrhosis and, in some cases, HCC (hepatocellular carcinoma) (9). In humans, cirrhotic patients showed impairment in SAMe biosynthesis as a result of decreased expression of *MAT1A* and MAT hepatic activity (10), resulting on low levels of GSH (glutathione). After SAMe administration, the level of GSH was increased (11), as well as the survival of patients (12).

However, not only the deficiency in hepatic SAMe biosynthesis leads to liver injury, but also abnormally increased methionine and SAMe contents. Patients with mutated *GNMT* (glycine *N*-methyltransferase) gene, which encodes the main methyltransferase responsible for the catabolism of SAMe excess (13), also develop steatosis, fibrosis and HCC (14). Moreover, a *GNMT* polymorphism (1289C > T) has been associated with HCC (15).

These findings regarding the importance of maintaining hepatic SAMe content within a specific range, are supported by those observed in mice lacking MAT1A and GNMT, with low and high levels of SAMe, respectively. Both groups of knockout mice successfully mimic the underlying pathologies observed in humans, spontaneously developing NASH and finally HCC, and in the case of GNMT−/− mice, fibrosis as well (14, 16). Although the expression of the pathology slightly differs among both groups of mice, a high susceptibility to damage induced by hepatotoxic agents and an impaired liver regeneration after partial hepatectomy are shared (17–19) supporting the key role of SAMe in the normal function of the liver.

## 2. Role of SAMe in Hepatocyte Proliferation

As the myth of Prometheus revealed, the high proliferative capacity of the liver after an injury is known since ancient Greeks times (20). In the bench, the best experimental model for the study of liver regeneration is partial hepatectomy (PH), the surgical removal of two-thirds of the liver tissue reported in rats by Higgins and Anderson (21). After PH, the regeneration of the liver is carried out by proliferation of the mature parenchymal cell populations, including hepatocytes, which divide once or twice to restore the original organ mass, without stem cell involvement (20, 22), a fact that raises this technique to the most often used stimulus to study the mechanisms involved in mature hepatocyte proliferation in vivo.

A large number of genes are involved in the complex cellular response, which allows liver regeneration, including cytokines and growth and metabolic factors. Among implicated hepatic mitogenic factors, one of the most important is HGF, secreted by mesenchymal cells. In rats, within 1 h after PH, the plasma level of HGF is increased more than 20-fold (23) and about 2 h later iNOS (inducible nitric oxide synthase) expression and NO (nitric oxide) biosynthesis are induced (24). NO reduces SAMe levels by specifically inhibiting MAT through $S$-nitrosylation of cysteine residue121 (25–27). These events lead hepatocytes into DNA synthesis and induction of early-response genes, as the first step for regeneration (27, 28).

Accumulating evidence indicates a key role of SAMe in hepatic growth. The administration of exogenous SAMe after PH prevents the reduction on cellular SAMe content and the DNA synthesis is inhibited, blocking the progression of regeneration (5). In rats previously treated with hepatocarcinogen, SAMe supplementation prevents the development of HCC (29) and, in cultured hepatocytes, is able to block the mitogenic effect of HGF (30). In addition, in MAT1A knockout mice, SAMe deficiency leads to uncontrolled hepatocyte proliferation (16), whereas in GNMT deficient mice, elevated levels of hepatic SAMe causes impaired liver regeneration after PH (19). The mechanism underlying the effect of SAMe content variation in the proliferation of hepatocytes was elucidated in the last years.

The proliferative response promoted by HGF includes the activation of four signaling pathways; Ras/ ERK (extracellular signal-regulated kinase)/MAPK (mitogen-activated protein kinase), PI3K/Akt, Rac/Pak and Crk/Rap1 (30, 31). Recently, the activation of an alternative noncanonical LKB1/AMPK (AMP-activated protein kinase)/eNOS (endothelial nitric oxide synthase) cascade has been reported (32). AMPK, a Serine/Threonine kinase involved in responding to cellular stresses by inhibiting cellular processes that consume energy and activating catabolic pathways that generate ATP (33), was also reported to regulate the subcellular localization of HuR (Human antigen R), an ubiquitously expressed RNA-binding protein (RBP) that increases the stability of target mRNAs (34). In hepatocytes, the activation of AMPK after HGF stimulus promotes the translocation from the nucleus to the cytosol of HuR, which stabilizes several cell-cycle genes such as cyclin A2 and D1, allowing the hepatocytes to proliferate (35). SAMe is able to prevent this process by inhibiting the phosphorylation of AMPK through a mechanism that likely involves the methylation of PP2A (protein phosphatase 2A) and its association to AMPK (35). In consequence, HuR is not transported to the cytoplasm and the stabilization of target genes does not occur, blocking HGF-induced proliferative response. A schematic representation of SAMe-regulated LKB1/AMPK/eNOS cascade and HuR involvement in HGF-induced hepatic growth is shown in Fig. 2.

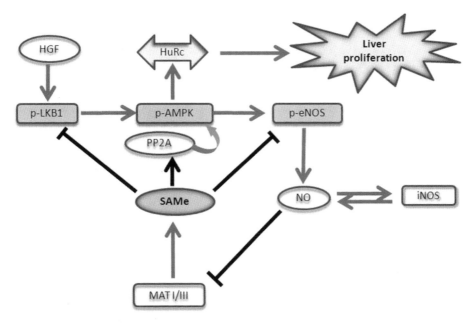

Fig. 2. Outline of the LKB1/AMPK/eNOS cascade. HGF induces the phosphorylation and activation of LKB1, AMPK and eNOS. AMPK phosphorylation induces the translocation to cytoplasm of HuR (HuRc), an RNA-binding protein that induces cell-cycle progression and hepatocyte proliferation by increasing the half-life of target mRNAs such as cyclin A2 and cyclin D1. eNOS-dependent NO production activates iNOS induction, which further contributes to NO synthesis, and methionine adenosyltransferase I/III (MAT I/III) inactivation, the main enzyme responsible for hepatic SAMe synthesis. This reduction in hepatic SAMe synthesis prevents the methylation and activation of PP2A, which further stimulates the phosphorylation and activation of the LKB1/AMPK/eNOS cascade and hepatocyte proliferation. SAMe treatment inhibits HGF-mediated hepatocyte proliferation via stimulation of PP2A methylation and inactivation of LKB1 and AMPK.

## 3. Human Antigen R

In recent years, messenger RNA (mRNA) turnover has emerged as a posttranscriptional mechanism that critically contributes to regulate the gene expression pattern during different cellular processes (36–38). One of the major actors in this scenario is HuR, the ubiquitously expressed member of the *Hu/Elav* family of RBPs (39). All family members contain three RNA recognition motifs (RRM) with high affinity for adenines and uracils-rich elements (AU-rich elements or AREs) (40), usually found in the 3′ untranslated region (UTR) of labile mRNAs (41). As consequence of this selective binding, HuR stabilizes target mRNAs, increasing their half-life and/or translation. Other identified RBPs promote labile mRNA decay, such as AUF1, TTP (Tristetraprolin) and KSRP (KH-type splicing regulatory protein) (42).

The regulation of HuR function, like other RBPs, appears to be closely linked to its subcellular localization (43). HuR is predominantly (>90%) nuclear, but can be exported to the cytoplasm in response to a variety of agents, notably proliferative and stressful signals (44, 45), by a process that involves a nucleo-cytoplasmic shuttling domain (HNS), located in the hinge region between RRMs 2 and 3 (46), and its association with transport receptors such as CRM1 and Transportins 1 and 2 (TRN1 and 2) (47, 48). Although not fully resolved, HuR may initially bind target mRNAs in the nucleus and transport them into the cytoplasm, avoiding mRNA decay by inhibiting the de-adenylation and targeting to the exosome of the transcripts or by competing for binding sites with destabilizing RBPs that recognize AREs (37).

Among HuR target mRNAs described to date, are those encoding genes implicated in cellular processes such as proliferation (32), apoptosis (49) or inflammation (50). Regarding cellular differentiation process, HuR is known to have a putative role in the differentiation of specific cellular lineages, such as spermatocytes (51), myocytes (52, 53) and adipocytes (54). These studies show a predominantly nuclear localization of HuR in actively proliferating, undifferentiated cells, but strikingly abundant in the cytoplasm upon induction and duration of differentiation, returning to a nuclear presence upon the completion of cell differentiation process.

The precise mechanism underlying the stabilization of labile mRNAs by HuR is still not well understood. However, it was suggested that posttranslational modification of HuR critically influences this process (55). Up to now, this RBP was reported to be posttranslationally regulated by phosphorylation at different serine/threonine residues by protein kinases such as cell-cycle checkpoint kinase 2 (Chk2) (56) and cyclin dependent kinase 1 (Cdk1) (57) and by methylation by coactivator-associated arginine methyltransferase 1 (CARM1 or PRMT4), which specifically transfers the methyl group from SAMe to arginine 217 residue in HuR protein (58). Although the exact effect each posttranslational modification has in HuR behavior has to be investigated individually, in general terms they affect HuR's mRNA-stabilizing activity by altering its subcellular levels and/or influencing its ability to bind RNA.

## 4. SAMe and Hepatic Differentiation

In fetal liver, *MAT2A* expression predominates and is progressively replaced by *MAT1A* expression throughout liver development, reaching a minimum in the adult hepatocyte, the opposite process occurs during malignant transformation (Fig. 3) (59, 60). Consequently, and because of differences in MAT isozymes properties (4–6), the level of SAMe is higher in mature hepatocytes than in not completely differentiated cells, which suggest a role of SAMe in the developmental process of the liver.

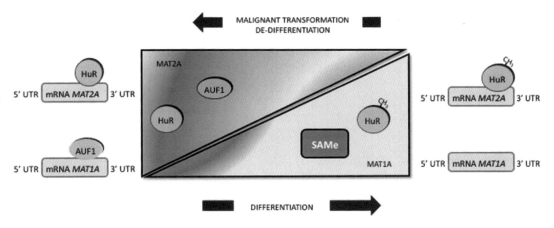

Fig. 3. Schematic representation of *MAT2A*, *MAT1A* and their posttranslational regulators HuR, methylated-HuR and AUF1 behavior during development, de-differentiation and malignant transformation of hepatocytes. Mature hepatocytes have high expression levels of *MAT1A* and low *AUF1* levels, while *MAT2A* is present in only very limited amounts due to the activity of its negative regulator, methylated-HuR. During the de-differentiation process, the levels of *AUF1* mRNA in hepatocytes increase, the ratio of methylated-HuR/HuR decrease, and as a consequence, there is a switch from *MAT1A* to *MAT2A* mRNA production. SAMe treatment of hepatocytes prevents these changes and maintains consistent levels of the RNA-binding proteins and *MAT1A* expression. During malignant transformation of the hepatocytes, a similar pattern of expression of *AUF1*, *HuR* and methylated-HuR, and *MAT1A* and *MAT2A* are observed. During liver development, the opposite is observed, and there is a decrease in *AUF1* levels and an increase in the methylated-HuR/HuR ratio.

It is known that primary hepatocytes de-differentiate, lose their hepatic polarity upon isolation and rapidly decline in liver-specific functions over culture time. At the same time, *MAT1A* expression progressively decreases, while *MAT2A* expression is induced (61). This switch is prevented by exogenous administration of SAMe to the culture medium, maintaining for longer time the differentiated status of the hepatocytes (61), through a mechanism that likely involves delaying the culture-induced expression of proteins expressed by early progenitor cells in the liver, such as Cx43 (conexin 43), and preventing the culture-induced decline of proteins related to mature hepatocytes such as Cx32 (conexin 32) and albumin (unpublished data). These findings identify SAMe as a molecule that differentially regulates gene expression and helps to maintain the functional and differentiated stage of the liver. This concept is further supported by the chemopreventive effects of SAMe exogenous administration on the development of preneoplastic lesions and HCC in models of rat liver carcinogenesis (62).

Studies recently assessed in proliferative and de-differentiated rat hepatocytes, as well as in human HCC samples, demonstrated that the switch in *MAT1A/MAT2A* expression is due to the posttranscriptional regulation executed on *MAT1A* and *MAT2A* by AUF1, HuR and methyl-HuR (59). In poorly differentiated hepatocytes HuR associates with the *MAT2A* 3′ UTR, enhancing its mRNA stability and steady-state levels, whereas AUF1 associates with the MAT1A 3′ UTR decreasing its mRNA stability and steady-state abundance. On the other hand, in well-differentiated hepatocytes, methyl-HuR was described for the first time as a destabilizer of *MAT2A* mRNA (59) (Fig. 3).

As happens in hepatic proliferation, this SAMe/HuR feedback regulation activity could be done through a mechanism involving the activation of AMPK. However, other signaling pathways have been reported to control nucleo-cytoplasmic shuttling of HuR, such as p38-MAPK, MAPKAPK-2 (MK2) and protein kinase c (PKC) (49), and the list is steadily growing. Moreover, SAMe, as the main cellular methyl group donor, is able to donate its methyl group to a large variety of acceptor molecules including DNA and RNA nucleic acids, phospholipids and proteins, greatly expanding the range of possibilities.

## 5. Usefulness of In Vitro Hepatic Differentiation Models to Elucidate the Regulatory Role of SAMe and HuR in Hepatocyte Differentiation

In order to uncover the underlying mechanisms by which SAMe level governs the hepatic differentiating program, an exhaustive study in hepatocytes at different maturation stages should be done. However, the difficulty of in vivo carrying out of some of the necessary experiments to clarify the proposed approach, such as silencing or overexpressing the hypothetically involved molecule, makes essential to find an appropriate in vitro model. The most important advantage of working with in vitro culture systems is the possibility to chemically interfere with the cellular process of interest and study the consequences, which lead us to refute or rule out the proposed hypothesis. Owing to the fact that HGF is a potent mitogen not only in vivo but also for hepatocytes in culture (20), the process of hepatic proliferation induced after PH can be easily reproduced in vitro by the addition of this growth factor in the culture medium, which has meant an important step forward in the study of liver regeneration.

Regarding the hepatic developmental process, recapitulating in culture the program in vivo seems to be more difficult. Although the implication of several pathways has been elucidated (63, 64), the overall understanding of their synchronicity and complementarity is still limited, mainly because of the lack of a suitable model. In the last years, this biological event was tried to be efficiently reproduced by priming embryonic stem cells (65, 66), mesenchymal stem cells (67–69), liver progenitor cells (70–72) or, more recently, induced-pluripotent stem cells (73, 74) towards a hepatocyte lineage and/or complete hepatocyte maturation by different combinations of growth factors and matrixes. However, most methods resulted in a heterogeneous cell population or low yields of cells of interest, which leads to dismiss them as reliable models. The interest in defining an efficient device that enables the specific hepatic fate of stem/progenitor cells is based not only on the usefulness for studying the molecular basis of hepatocyte differentiation per se but also on its potential to provide a continual source of liver cells for therapeutic and pharmaco-toxicological purposes (75). In consequence, references in this field increase and are renewed constantly. Although to date the description of a highly productive and standardized experimental protocol is still a challenge, recently, as show several chapters in this book, improvements in the culture conditions have been reported to produce a higher purity of functional hepatocyte-like cells (76). Therefore, new insights on the understanding of how hepatogenesis is controlled can be expected to appear in the near future.

## 6. General Methodology to Uncover Signaling Cascades with SAMe as Key Regulator

The usual methodology to unmasking the mechanism that modulates a specific cellular program starts with the identification and analysis of the proteins that are part of the signaling pathway triggered or regulated by specific physiological conditions or by the action of a concrete stimulus. According to the hypothesis argued in this chapter, SAMe would act as a regulatory agent of the hepatic developmental process. Therefore, interfering the cellular SAMe content should cause changes in the intracellular signaling cascade, in turn modifying the normal differentiation program. Those changes are key clues to elucidate the constituents of the signaling pathway of interest. After identifying those proteins, the hypothetical tangle of intracellular signaling events must be validated. The usual method consists on reducing or increasing the function of a chosen protein by gene silencing and gene overexpression, respectively, and checking how this alteration affects the subsequent downstream cascade by using techniques such as western blotting. Moreover, considering that SAMe acts as a regulator of HuR functionality, which entails its translocation to the cytosol and stabilization of target mRNAs, a verification of HuR location and functionality would be necessary. Furthermore, taking into consideration that the products of HuR target mRNAs are finally the executors of the cellular response, the identification of those targets in each stage of the hepatic developmental process would help to widen the understanding of this complex program and to complete the succession of intracellular reactions which make the cell capable of differentiating.

The different techniques that make up this methodology are briefly described in the sections below.

### 6.1. How to Alter Cellular SAMe Content In Vitro

There are two methods to mimic the depletion of cellular SAMe level causes on normal cell behavior. One method involves culturing the cells in culture medium without methionine, precursor of SAMe (77). The other method uses a reagent of analytical grade obtained from commercial sources, cycloleucine. Cycloleucine (1-aminocyclopentane-1-carboxylic acid) is a non-metabolizable synthetic amino acid that competitively inhibits MAT resulting in the blockade of SAMe synthesis (78). Its addition to the culture medium has been reported to cause neither cytotoxicity nor loss of cell viability to primary hepatocytes (79), two key points to take into account when treating cells with synthetic agents. The dosage to use depends on the percentage of SAMe depletion required for the study. A high concentration of cycloleucine, 20 mM, in the culture medium produces an 80% loss of SAMe content 24 h after the treatment (78), while a concentration of 5 mM results in an approximate 50% fall (79). On the other hand, the consequences the increase in cellular SAMe content has in the succession of intracellular events of interest are evaluated by the addition of commercially available SAMe to the culture medium at a high final concentration of 4 mM, due to its low permeability (35). In both cases, to verify the success of the treatment, cellular SAMe content is analyzed by LC/MS (liquid chromatography/ mass spectrometry) using an ACQUITY-UPLC system coupled to a LCT Premier Mass Spectrometer equipped with a spray ionization source (14, 61).

### 6.2. Western/Protein Blotting to Detect HuR and Other Proteins

This technique allows the detection, measurement and characterization of certain proteins from a wide range of sample types and the comparison for protein content between samples from different origins (80). The protein sample to be analyzed can be obtained both from whole-cell lysates and from subcellular fractions. Therefore, this method not only enables to detect the global effects of cellular SAMe content at protein level but also determine the subcellular localization of HuR (35). Proteins are eletrophoretically separated using an SDS polyacrylamide gel according to their molecular weight and subsequently transferred and immobilized on Polyvinylidene Fluoride (PVDF) or nitrocellulose membranes before detection by using specific ligands such as polyclonal and monoclonal antibodies. Among the wide range of commercially available antibodies, some of them have been developed against the consensus phosphorylated peptide sequences of specific kinases whose activation depends on phosphorylation, such as AMPK (32, 35), in order to provide a precise measure of enzyme activation. Then, blots are developed by enhanced chemoluminiscence and exposed to x-ray film. Finally, the semi-quantitative analysis of protein level between samples is evaluated through densitometry of blots, which requires digitalization of x-ray films and translation of protein bands intensity in the resulting images into values that subsequently can be presented as graphs (81).

### 6.3. Immunofluorescence for HuR Subcellular Detection

Another method to visualize the subcellular localization of a protein, in our case, endogenous HuR is performing immunofluorescence assay (35). Cells, plated and cultured onto round cover slips, are fixed in cold methanol followed by blocking of the unspecific binding sites and membrane permeabilization. Next, cells are incubated with the specific antibody against HuR and then with the corresponding secondary antibody conjugated with a fluorochrome such as FITC (fluorescein isothiocyanate). Nuclei visualization is done by using fluorescent dyes that selectively bind to double stranded DNA, usually DAPI (diamidino-2-phenylindole) and Hoechst 33342, the latter with a lower photostability. Finally, the cell samples are mounted using commercially available mounting solutions and immune complexes detected by fluorescence or confocal microscopy at an excitation wavelength dependent on the chosen fluorochrome.

### 6.4. How to Modify Gene Expression

Once the proteins that make up the signaling pathways regulated by SAMe have been identified, the precise flux of intracellular events must be checked and verified. In the last decades, significant advances have been made in understanding the fundamental principles that govern the process by which a gene is translated into a functional protein. Consequently, the ability of a gene to express biologically active proteins can be controlled on the bench.

#### 6.4.1. Gene Silencing

Gene silencing is the term used to describe the mechanism to inhibit the expression of a gene. In our studies, we used the RNAi (RNA interference) technique, which has been demonstrated as a valuable method for posttranscriptional silencing studies (82), to repress at the mRNA level the expression of the proteins LKB1, AMPK and eNOS (32) and HuR (59) in order to uncover the molecular pathway controlled by SAMe in hepatic proliferation and differentiation processes, respectively. We selected the hepatocyte cell line (MLP29), plated 24 h before, to be transfected with the exogenous siRNA (small interfering RNA) using a cationic liposome-based reagent that provides high transfection efficiency and high levels of transgene expression. 24–72 h later, the target mRNA knockdown level is checked by quantitative reverse transcription-polymerase chain reaction (RT-PCR) and the protein knockdown level verified by western/protein blotting. In each experiment, one of the members of the signaling cascade of study was inhibited and the downstream events analyzed by western blotting and compared to those observed in negative control samples.

*6.4.2. Gene*
*Overexpression*

Gene overexpression is the term used to describe the technique developed to increase the expression of a gene. The usual method followed for this purpose relies on the fact that exogenous DNA containing fragments of a specific gene is randomly inserted into the genome of the transfected cells and transiently expressed, increasing the expression level of that gene and, consequently, its corresponding functional protein. We followed this method to overexpress the RNA-binding recognition domain of HuR, the nucleotide sequence responsible for its function (59). For that, hepatic cells were transfected with a DNA plasmid containing the selected nucleotide sequence and GFP (green fluorescent protein) gene by using the same liposome-based transfection reagent mentioned above. After 4–16 h the complexes were removed. The overexpression of the targeted gene and protein is checked by quantitative RT-PCR and western blotting, respectively. The effects in the downstream intracellular events were also checked by immunoblotting.

**6.5. Chemical**
**Inhibitors**
**and Activators**

Apart from RNAi, another way to get the functional inactivation of a single gene to clarify the downstream events is the use of specific cell-permeant chemical inhibitors. The key point of this technique is to achieve a selective inhibition, without causing toxicity or alteration in the cell viability and without affecting the normal functionality of secondary molecules, which could tangle the biological response to the drug. Therefore, it is necessary to choose an inhibitor with a well-understood role and competitive inhibitory effect. Nowadays, many inhibitors available on the market have been extensively used and reported as highly specific drugs, such as the MEK inhibitor PD098059 (83, 84) or PI3kinase inhibitor LY294002 (85, 86), arising as extremely powerful tools for analyzing signal transduction processes.

The opposite effect, the activation of a target enzyme, is also easily reproducible in vitro by using specific activators. For instance, in our study focused on elucidating the role of AMPK phosphorylation in the translocation of HuR from the nucleus to the cytosol, for AMPK specific activation we used AICAR (5-aminoimidazole-4-carboxamideriboside), the most widely used pharmacologic activator of this kinase, at a final concentration of 2 mM (35).

**6.6. Detection of RNA**
**Bound to HuR**
**by RNP-Ip**

RNP-Ip is the method which enables the immunoprecipitation of HuR using the antigen–antibody reaction principle and the collection of subsets of mRNAs bound to it for further identification by genomic array technology or RT-qPCR (87). Apart from being useful to identify HuR target mRNAs, this method complements the information obtained by performing the immunofluorescence assay, allowing us to verify the functionality of HuR in the cell system. In our studies, for endogenous mRNA-HuR complexes isolation, the whole cell lysate is incubated with Protein A-Sepharose beads previously coated with a specific antibody against HuR and treated with RNase-free DNAse I. After phenol extraction, RNA is precipitated with ethanol containing a blue dye covalently linked to glycogen that coprecipitates with the RNA to facilitate its visualization and recovery. Finally, the collected subsets of mRNA are retrotranscribed for further identification by genomic array technology or, in case of an already identified target mRNA, for expression analysis by quantitative PCR (35, 59).

***6.7. Validation/ Confirmation of HuR target mRNA: Biotin Pull Down***

This in vitro technique permits to validate the interaction between a specific mRNA sequence and an RBP, such as HuR, as well as to know the specific sequence region where the binding occurs. Our team carried out this method to test the possibility that *MAT2A* and *MAT1A* mRNAs were targets of HuR and AUF1, respectively (59). It starts with the synthesis of biotinylated transcripts corresponding to the mRNA of interest. For that purpose, total RNA of cultured hepatic cells is collected, reverse-transcribed into cDNA and amplified by PCR using specific primers for different overlap fragments of the target genes, called probes, and 5′ oligonucleotides containing the T7 RNA polymerase promoter sequence. Next, the PCR products are purified and used as template for the synthesis of the corresponding biotinylated RNA using T7 RNA polymerase and biotin-CTP (cytidine triphosphate). Then, the biotinylated probe is purified and incubated with the protein cell lysate to analyze in the presence of RNase inhibitor. Finally, RNA-HuRcomplexes are isolated using streptavidin-conjugated magnetic beads and analyzed by western blotting using a specific antibody that recognizes HuR. If the probe interacts with HuR it might be in the complex beads-probe-protein and, therefore, detected in the blot.

# 7. Summary and Conclusions

Hepatic differentiation is a complex process that requires the balanced regulation of multiple pathways. Although significant advances have been made in understanding the molecular mechanisms that modulate the onset of hepatogenesis, the overall understanding is still unclear. SAMe, the main methyl group donor in the cell, since its discovery, has emerged as a key molecule that plays a central role in numerous hepatic processes. Its regulatory effect in the proliferative response of hepatocytes involves the activation of LKB1/AMPK/eNOS cascade and HuR function. Recently, SAMe and HuR were also reported to execute a modulation on the hepatic differentiation program. In order to deeply examine the functioning and regulation of both molecules and unravel the signaling pathways implicated, the utilization of in vitro models that can reproduce physiological events related to hepatocyte differentiation is worthwhile. Although a highly effective and reproducible culture system that triggers stem/progenitor cells into functional hepatocytes is still a challenge, a satisfactory progress in this field has been made in the last years, as show several chapters included in this book, representing an attractive approach for studying the implication of SAMe and HuR in liver development.

## Acknowledgments

This study was supported by AT-1576 (to JMM and MLM–C), SAF2005-00855, HEPADIP-EULSHM-CT-205, and ETORTEK-2008 (to JMM and MLM–C); Program Ramón y Cajal del MEC and Fundación "La Caixa" (to MLM–C); and Centro de Investigación Biomédica en Red de Enfermedades Hepáticas y Digestivas is funded by the Instituto de Salud Carlos III.

## References

1. Cantoni G.L. (1975) Biological methylation: selected aspects. *Annu Rev Biochem* **44**, 435–51.

2. Finkelstein J.D., Martin J.J. (1984) Methionine metabolism in mammals. Distribution of homocysteine between competing pathways. *J Biol Chem* **259**, 9508–9513.

3. Mato J.M., Alvarez L., Ortiz P. et al. (1997) S-adenosylmethionine synthesis: molecular mechanisms and clinical implications. *Pharmacol Ther* **73**, 265–280.

4. Mato J.M., Corrales F.J., Lu S.C. et al. (2002) S-adenosylmethionine: a control switch that regulates liver function. *FASEB J* **16**, 15–26.

5. Mato J.M., Lu S.C. (2007) Role of S-adenosyl-l-methionine in liver health and injury. *Hepatol* **45**, 1306–1312.

6. Cai J., Mao Z., Hwang J.J. et al. (1998) Differential expression of methionine adenosyltransferase genes influences the rate of growth of human hepatocellular carcinoma cells. *Cancer Res* **58**, 1444–1450.

7. Mudd S.H., Poole J.R. (1975) Labile methyl balances for normal humans on various dietary regimens. *Metabol* **24**, 721–735.

8. Best C.H., Hershey J.M., Huntsman M.E. (1932) The effect of lecithin on fat deposition in the liver of the normal rat. *J Physiol* **75**, 56–66.

9. Shivapurkar N., Poirier L.A. (1983) Tissue levels of S-adenosylmethionine and S-adenosyl homocysteine in rats fed methyl-deficient, amino acid-defined diets for one to five weeks. *Carcinogenesis* **4**, 1051–157.

10. Duce A.M., Ortiz P., Cabrero C. et al. (1988) S-Adenosyl-L-methionine synthetase and phospholipid methyltransferase are inhibited in human cirrhosis. *Hepatol* **8**, 65–68.

11. Vendemiale G., Altomare E., Trizio T. et al. (1989) Effect of oral S-adenosyl-L-methionine on hepatic glutathione in patients with liver disease. *Stand J Gastroenterol* **24**, 407–415.

12. Mato J.M., Cámara J., Fernández de Paz J. et al. (1999) S-adenosylmethionine in alcoholic liver cirrhosis: a randomized, placebo-controlled, double-blind, multicenter clinical trial. *Hepatol* **30**, 1081–1089.

13. Mudd S.H., Brosnan J.T., Brosnan M.E. et al. (2007) Methyl balance and transmethylation fluxes in humans. *Am J Clin Nutr* **85**, 19–25.

14. Martínez-Chantar M.L., Vázquez-Chantada M., Ariz U. et al. (2008) Loss of the glycine N-methyltransferase gene leads to steatosis and hepatocellular carcinoma in mice. *Hepatol* **47**, 1191–1199.

15. Tseng T.L., Shih Y.P., Huang Y.C. et al. (2003) Genotypic and phenotypic characterization of a putative tumor susceptibility gene, GNMT, in liver cancer. *Cancer Res* **63**, 647–654.

16. Martinez-Chantar M.L., Corrales F.J., Martinez-Cruz L.A. et al. (2002) Spontaneous oxidative stress and liver tumors in mice lacking methionine adenosyltransferase 1A. *FASEB J* **16**, 1292–1294.

17. Lu S.C., Alvarez L., Huang Z.Z. et al. (2001) Methionine adenosyltransferase 1A knockout mice are predisposed to liver injury and exhibit increased expression of genes involved in proliferation. *Proc Natl Acad Sci USA* **98**, 5560–5565.

18. Chen L., Zeng Y., Yang H. et al. (2004) Impaired liver regeneration in mice lacking methionine adenosyltransferase 1A. *FASEB J* **18**, 914–916.

19. Varela-Rey M., Fernández-Ramos D., Martínez-López N. et al. (2009) Impaired liver regeneration in mice lacking glycine N-methyltransferase. *Hepatol* **50**, 443–452.

20. Michalopoulos G.K., DeFrances M.C. (1997) Liver regeneration. *Science* **276**, 60–66.

21. Higgins G.M., Anderson R.M. (1931) Experimental pathology of the liver. I. Restoration of the liver of the white rat following partial surgical removal. *Arch Pathol* **12**, 186–202.

22. Fausto N., Campbell J.S., Riehle K.J. (2006) Liver regeneration. *Hepatol* **43**(2 Suppl 1): S45–53.

23. Lindroos P.M., Zarnegar R., Michalopoulos G.K .(1991) Hepatocyte growth factor (hepatopoietin A) rapidly increases in plasma before DNA synthesis and liver regeneration stimulated by partial hepatectomy and carbon tetrachloride administration. *Hepatol* **13**, 743–750.

24. Hortelano S., Dewez B., Genaro A.M. et al. (1995) Nitric oxide is released in regenerating liver after partial hepatectomy. *Hepatol* **21**, 776–786.

25. Ruiz F., Corrales F.J., Miqueo C. et al. (1998) Nitric oxide inactivates rat hepatic methionine adenosyltransferase In vivo by S-nitrosylation. *Hepatol* **28**, 1051–1057.

26. Pérez-Mato I., Castro C., Ruiz F.A. et al. (1999) Methionine adenosyltransferase S-nitrosylation is regulated by the basic and acidic amino acids surrounding the target thiol. *J Biol Chem* **274**, 17075–17079.

27. Garcia-Trevijano E.R., Martinez-Chantar M.L., Latasa M.U. et al. (2002) NO sensitizes rat hepatocytes to proliferation by modifying S-adenosylmethionine levels. *Gastroenterol* **122**, 1355–1363.

28. Huang Z.Z., Mao Z., Cai J. et al. (1998) Changes in methionine adenosyltransferase during liver regeneration in the rat. *Am J Physiol* **275**, G14–G21.

29. Pascale R.M., Simile M.M., De Miglio M.R. et al. (1995) Chemoprevention by S-adenosyl-L-methionine of rat liver carcinogenesis initiated by 1,2-dimethylhydrazine and promoted by orotic acid. *Carcinogenesis* **16**. 427–430.

30. Ponzetto C., Bardelli A., Zhen Z. et al. (1994) A multifunctional docking site mediates signaling and transformation by the hepatocyte growth factor/scatter factor receptor family. *Cell* **77**, 261–277.

31. Paumelle R., Tulasne D., Kherrouche Z. et al. (2002) Hepatocyte growth factor/scatter factor activates the ETS1 transcription factor by a RAS-RAF-MEK-ERK signaling pathway. *Oncogene* **21**, 2309–2319.

32. Vázquez-Chantada M., Ariz U., Varela-Rey M. et al. (2009) Evidence for LKB1/AMP-activated protein kinase/endothelial nitric oxide synthase cascade regulated by hepatocyte growth factor, S-adenosylmethionine, and nitric oxide in hepatocyte proliferation. *Hepatol* **49**, 608–617.

33. Hardie D.G., Hawley S.A., Scott J.W. (2006) AMP-activated protein kinase-development of the energy sensory concept. *J Physiol* **574**, 7–15.

34. Wang W., Fan J., Yang X. et al. (2002) AMP-activated kinase regulates cytoplasmic HuR. *Mol Cell Biol* **22**, 3425–3436.

35. Martinez-Chantar M.L., Vazquez-Chantada M., Garnacho M. et al. (2006) S-adenosylmethionine regulates cytoplasmic HuR via AMP-activated kinase. *Gastroenterol* **131**, 223–232.

36. Bolognani F., Perrone-Bizzozero N.I. (2008) RNA-protein interactions and control of mRNA stability in neurons. *J Neurosci Res* **86**, 481–489.

37. Garneau N.L., Wilusz J., Wilusz C.J. (2007) The highways and byways of mRNA decay. *Nat Rev Mol Cell Biol* **8**, 113–126.

38. Wilusz C.J., Wormington M., Peltz S.W. (2001) The cap-to-tail guide to mRNA turnover. Nat Rev *Mol Cell Biol* **2**, 237–246.

39. Ma W.J., Cheng S., Campbell C. et al. (1996) Cloning and characterization of HuR, a ubiquitously expressed Elav-like protein. *J Biol Chem* **271**, 8144–8151.

40. Park S., Myszka D.G., Yu M. et al. (2000) HuD RNA recognition motifs play distinct roles in the formation of a stable complex with AU-rich RNA. *Mol Cell Biol* **20**, 4765–4772.

41. Xu N., Chen C.Y., Shyu A.B. (1997) Modulation of the fate of cytoplasmic mRNA by AU-rich elements: key sequence features controlling mRNA deadenylation and decay. *Mol Cell Biol* **17**, 4611–4621.

42. Bevilacqua A., Ceriani M.C., Capaccioli S. et al. (2003) Post-transcriptional regulation of gene expression by degradation of messenger RNAs. *J Cell Physiol* **195**, 356–372.

43. Keene J.D. (1999) Why is Hu where? Shuttling of early-response-gene messenger RNA subsets. *Proc Natl Acad Sci USA* **96**, 5–7.

44. Wang W., Furneaux H., Cheng H. et al. (2000) HuR regulates p21 mRNA stabilization by UV light. *Mol Cell Biol* **20**, 760–769.

45. Wang W., Caldwell M.C., Lin S. et al. (2000) HuR regulates cyclin A and cyclin B1 mRNA stability during cell proliferation. *EMBO J* **19**, 2340–2350.

46. Fan X.C., Steitz J.A. (1998) HNS, a nuclear-cytoplasmic shuttling sequence in HuR. *Proc Natl Acad Sci USA* **95**, 15293–15298.

47. Güttinger S., Mühlhäusser P., Koller-Eichhorn R. et al. (2004)Transportin2 functions as importin and mediates nuclear import of HuR. *Proc Natl Acad Sci USA* **101**, 2918–2923.

48. Rebane A., Aab A., Steitz J.A. (2004) Transportins 1 and 2 are redundant nuclear import factors for hnRNP A1 and HuR. *RNA* **10**, 590–599.

49. Abdelmohsen K., Lal A., Kim H.H. et al. (2007) Posttranscriptional orchestration of an anti-apoptotic program by HuR. *Cell Cycle* **6**, 1288–1292.

50. Katsanou V., Papadaki O., Milatos S. et al. (2005) HuR as a negative posttranscriptional modulator in inflammation. *Mol Cell* **19**, 777–789.

51. Levadoux-Martin M., Gouble A., Jégou B. et al. (2003) Impaired gametogenesis in mice that overexpress the RNA-binding protein HuR. *EMBO Rep* **4**, 394–399.

52. Van der Giessen K., Gallouzi I. E. (2007) Involvement of transportin 2-mediated HuR import in muscle cell differentiation. *Mol Biol Cell* **18**, 2619–2629.

53. Figueroa A., Cuadrado A., Fan J. et al. (2003) Role of HuR in skeletal myogenesis through coordinate regulation of muscle differentiation genes. *Mol Cell Biol* **23**, 4991–5004.

54. Gantt K., Cherry J., Tenney R. et al. (2005) An early event in adipogenesis, the nuclear selection of the CCAAT enhancer-binding protein {beta} (C/EBP{beta}) mRNA by HuR and its translocation to the cytosol. *J Biol Chem* **280**, 24768–2474.

55. Doller A., Pfeilschifter J., Eberhardt W. (2008) Signalling pathways regulating nucleo-cytoplasmic shuttling of the mRNA-binding protein HuR. *Cell Signal* **20**, 2165–2173.

56. Abdelmohsen K., Pullmann R Jr., Lal A. et al. (2007) Phosphorylation of HuR by Chk2 regulates SIRT1 expression. *Mol Cell* **25**, 543–557.

57. Kim H. H., Abdelmohsen K., Lal A. et al. (2008) Nuclear HuR accumulation through phosphorylation by Cdk1. *Genes Dev* **22**, 1804–1815.

58. Li H., Park S., Kilburn B. et al. (2002) Lipopolysaccharide-induced methylation of HuR, an mRNA-stabilizing protein, by CARM1. Coactivator-associated arginine methyltransferase. *J BiolChem* **277**, 44623–44630.

59. Vázquez-Chantada M., Fernández-Ramos D. et al. ( 2010) HuR/Methyl-HuR and AU-Rich RNA Binding Factor 1 Regulate the Methionine Adenosyltransferase Expressed During Liver Proliferation, Differentiation, and Carcinogenesis. *Gastroenterol* **138**, 1943–1953.

60. Cai J., Sun W.M., Hwang J.J. et al. (1996) Changes in S-adenosylmethionine synthetase in human liver cancer: molecular characterization and significance. *Hepatol* **24**, 1090–1097.

61. García-Trevijano E. R., Latasa M.U., Carretero M. V. et al. (2000) S-adenosylmethionine regulates MAT1A and MAT2A gene expression in cultured rat hepatocytes: a new role for S-adenosylmethionine in the maintenance of the differentiated status of the liver. *FASEB J* **14**, 2511–2518.

62. Pascale R.M., Marras V., Simile M.M. et al. (1992) Chemoprevention of rat liver carcinogenesis by S-adenosyl-L-methionine: a long-term study. *Cancer Res* **52**, 4979–4986.

63. Si-Tayeb K, Lemaigre FP, Duncan SA (2010) Organogenesis and development of the liver. *Dev Cell* **18**, 175–189.

64. Lemaigre F.P. (2009) Mechanisms of liver development: concepts for understanding liver disorders and design of novel therapies. *Gastroenterol* **137**, 62–79.

65. Yamamoto H., Quinn G., Asari A. et al. (2003) Differentiation of embryonic stem cells into hepatocytes: biological functions and therapeutic application. *Hepatol* **37**, 983–993

66. Cai J., Zhao Y., Liu Y. et al. (2007) Directed differentiation of human embryonic stem cells into functional hepatic cells. *Hepatol* **45**, 1229–1239.

67. Banas A., Teratani T., Yamamoto Y. et al. (2007) Adipose tissue-derived mesenchymal stem cells as a source of human hepatocytes. *Hepatol* **46**, 219–228.

68. Kang X.Q., Zang W.J., Song T.S. et al. (2005) Rat bone marrow mesenchymal stem cells differentiate into hepatocytes in vitro. *World J Gastroenterol* **11**, 3479-3484.

69. Snykers S., Vanhaecke T., De Becker A. et al. (2007) Chromatin remodeling agent trichostatin A: a key-factor in the hepatic differentiation of human mesenchymal stem cells derived of adult bone marrow. *BMC Dev Biol* **7**, 24.

70. Tanimizu N., Saito H., Mostov K. et al. (2004) Long-term culture of hepatic progenitors derived from mouse Dlk + hepatoblasts. *J Cell Sci* **117**(Pt 26), 6425–6434.

71. Fujikawa T., Hirose T., Fujii H. et al. (2003) Purification of adult hepatic progenitor cells using green fluorescent protein (GFP)-transgenic mice and fluorescence-activated cell sorting. *J Hepatol* **39**, 162–170.

72. Nowak G., Ericzon B.G., Nava S. et al. (2005) Identification of expandable human hepatic progenitors which differentiate into mature hepatic cells in vivo. *Gut* **54**, 972–979.

73. Si-Tayeb K., Noto F.K., Nagaoka M. et al. (2010) Highly efficient generation of human hepatocyte-like cells from induced pluripotent stem cells. *Hepatol* **51**, 297–305.

74. Gai H, Leung EL, Costantino PD et al (2009) Generation and characterization of functional cardiomyocytes using induced pluripotent stem cells derived from human fibroblasts. *Cell Biol Int* **33**, 1184–1193.

75. Lysy P.A., Campard D., Smets F. et al. (2008) Stem cells for liver tissue repair: current knowledge and perspectives. *World J Gastroenterol* **14**, 864–875.

76. Cho C.H., Parashurama N., Park E.Y. et al. (2008) Homogeneous differentiation of hepatocyte-like cells from embryonic stem cells: applications for the treatment of liver failure. *FASEB J* **22**, 898–909.

77. Martínez-Chantar M.L., Latasa M.U., Varela-Rey M. et al. (2003) L-methionine availability regulates expression of the methionine adenosyltransferase 2A gene in human hepatocarcinoma cells: role of S-adenosylmethionine. *J Biol Chem* **278**, 19885–19890.

78. Yang H., Sadda M.R., Li M. et al. (2004) S-adenosylmethionine and its metabolite induce apoptosis in HepG2 cells: Role of protein phosphatase 1 and Bcl-x(S). *Hepatol* **40**, 221–231.

79. Zhuge J (2008) A decrease in S-adenosyl-L-methionine potentiates arachidonic acid cytotoxicity in primary rat hepatocytes enriched in CYP2E1. *Mol Cell Biochem* **314**, 105–112.

80. Heermann K.H., Gültekin H., Gerlich W.H. (1988) Protein blotting: techniques and application in virus hepatitis research. *Ric Clin Lab* **18**, 193–221.

81. Gassmann M., Grenacher B., Rohde B. et al. (2009) Quantifying Western blots: pitfalls of densitometry. *Electrophoresis* **30**, 1845–1855.

82. Marx J. (2000) Interfering with gene expression. *Science* **288**, 1370–1372.

83. Alessi D.R., Cuenda A., Cohen P. et al. (1995) PD 098059 is a specific inhibitor of the activation of mitogen-activated protein kinase kinase in vitro and in vivo. *J Biol Chem* **270**, 27489–27494.

84. Fehrenbach H., Weiskirchen R., Kasper M. et al. (2001) Up-regulated expression of the receptor for advanced glycation end products in cultured rat hepatic stellate cells during transdifferentiation to myofibroblasts. *Hepatol* **34**, 943–952.

85. Vlahos C.J., Matter W.F., Hui K.Y. et al. (1994) A specific inhibitor of phosphatidylinositol 3-kinase, 2-(4-morpholinyl)-8-phenyl-4H-1-benzopyran-4-one (LY294002). *J Biol Chem* **269**, 5241–5248.

86. Duran J., Obach M., Navarro-Sabate A. et al. (2009) Pfkfb3 is transcriptionally upregulated in diabetic mouse liver through proliferative signals. *FEBS J* **276**, 4555–4568.

87. Tenenbaum S.A., Carson C.C., Lager P.J. et al. (2000) Identifying mRNA subsets in messenger ribonucleoprotein complexes by using cDNA arrays. *Proc Natl Acad Sci USA* **97**, 14085–14090.

# Part III

## BD Formation from Stem Cells

# Chapter 13

# Transdifferentiation of Mature Hepatocytes into Bile Duct/ductule Cells Within a Collagen Gel Matrix

## Yuji Nishikawa

## Abstract

The phenotype of hepatocytes has been thought to be fixed once they are terminally differentiated. However, we and other investigators have demonstrated that mature hepatocytes can transform into bile duct/ductule cells in various experimental conditions in vitro. Since the normal bile duct system is almost invariably surrounded by dense periportal collagenous matrices, we placed isolated hepatocytes in a collagen-rich environment to address whether mature hepatocytes can transform into ductular cells. Here, we describe in detail our three-dimensional collagen culture method for the induction of transdifferentiation of mature rat hepatocytes into bile ductular cells. Our in vitro system might be useful for the elucidation of the mechanisms of the aberrant differentiation of hepatocytes in the diseased liver.

**Key words:** Hepatocytes, Bile duct cells, Liver fibrosis, Transdifferentiation, Inflammatory cytokines, Extracellular matrices, Three-dimensional cultures

## 1. Introduction

An increase of irregular ductular structures has been frequently observed in various liver diseases associated with fibrosis (Fig. 1a) (1, 2). This phenomenon called atypical ductular reaction has been ascribed to be the result of regenerative proliferation of the so-called liver stem cells, which are supposed to reside in the periportal region, particularly in the canal of Hering (3). The stem cell proliferation has been considered to take place when regenerative proliferation of parenchymal cells is compromised by liver injury (4). Mature hepatocytes have been shown to have a stem cell-like proliferating potential, but have been regarded not to have a capacity to differentiate into other cells, including bile duct/ductule cells (5). However, the possibility of ductular metaplasia of hepatocytes,

Takahiro Ochiya (ed.), *Liver Stem Cells: Methods and Protocols*, Methods in Molecular Biology, vol. 826,
DOI 10.1007/978-1-61779-468-1_13, © Springer Science+Business Media, LLC 2012

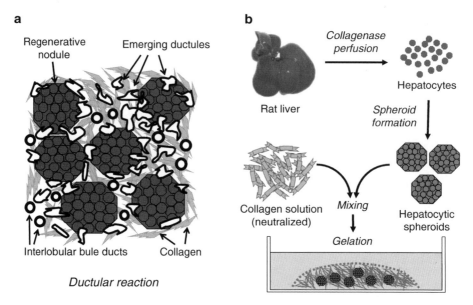

Fig. 1. Schematic representation of ductular reaction and the three-dimensional model for ductular transdifferentiation of rat hepatocytes. (**a**) Ductular reaction. Chronic liver injury is associated with hepatocyte damage and subsequent regeneration (regenerative nodules), inflammation, and reparative tissue reaction (fibrosis). Emergence of ductular structures with irregular contours (ductular reaction) takes place at the periphery of regenerative nodules where type I collagen is accumulated as the extracellular matrix. Interlobular bile ducts often remain intact. (**b**) Our three-dimensional model for hepatocytic transdifferentiation. Isolated rat hepatocytes are first cultured on Primaria dishes to form aggregates, and then, they are embedded within a collagen gel matrix. This model reproduces an analogous condition in which hepatocytes are placed in chronic liver injury.

especially in primary biliary cirrhosis, has also been proposed from the histopathological observations (6, 7).

Both hepatocytes and intrahepatic bile ducts/ductules are derived from hepatoblasts which are induced from the foregut endoderm (8). Importantly, the initiation of ductular differentiation is closely associated with the development of portal connective tissue (9). While bile ducts/ductules are surrounded by collagenous connective tissues, hepatocytes reside in the microenvironment which is typically very few in extracellular matrices (the space of Disse).

To address the phenotypic plasticity of mature hepatocytes, we placed spheroidal aggregates of primary rat hepatocytes in a three-dimensional collagen gel matrix (Fig. 1b) (10, 11). Our results demonstrated that hepatocytes transformed into bile ductular cells, without acquiring liver stem cell phenotypes, such as the expression of delta-like or α-fetoprotein, suggesting transdifferentiation, rather than dedifferentiation (11). Similar bile ductular differentiation of mature hepatocytes has also been demonstrated by other investigators in a three-dimensional culture system using roller bottles (12). Although there is a paucity of data showing that such transdifferentiation actually takes place in vivo (13), these observations might be potentially important in further understanding of the nature of ductular reaction in chronic liver diseases.

# 2. Materials

### 2.1. Isolation of Rat Hepatocytes

1. Adult male rats (10–20 weeks old). Any major strains, such as Fischer 344, Wistar, and Sprague–Dawley, can be used.

2. Preperfusion solution (Solution I): 9.52 g Hanks' balanced salt solution (Sigma), 0.19 g EGTA, 2.38 g Hepes, and 0.35 g NaHCO$_3$ are dissolved in distilled and deionized water (make up to 1 L, adjust to pH 7.2, and filter through a 0.22-μm Corning filter).

3. Perfusion solution (Solution II): 8 g NaCl, 0.4 g KCl, 0.56 g CaCl$_2$, 0.078 g NaH$_2$PO$_4$·2H$_2$O, 0.151 g Na$_2$HPO$_4$·12H$_2$O, 2.38 g Hepes, 6 mg phenol red, and 0.35 g NaHCO$_3$ are dissolved in distilled and deionized water (make up to 1 L, adjust to pH 7.35, and filter through a 0.22-μm Corning filter).

4. Collagenase solution: Dissolve 50 mg collagenase (e.g., Collagenase S-1 (Nitta gelatin, Osaka, Japan)] in 100 mL Solution II (final concentration, 0.05%) and filter through a 0.22-μm Corning filter (make fresh as required) (see Note 1).

5. Washing solution: 9.52 g Hanks' balanced salt solution, 0.35 g NaHCO$_3$, 0.14 g CaCl$_2$, MgCl$_2$·6H$_2$O, MgSO$_4$·7H$_2$O are dissolved in distilled and deionized water (make up to 1 L, adjust to pH 7.2, and filter through a 0.22-μm Corning filter).

6. 90% Percoll solution: Mix 90 mL of Percoll and 10 mL of 10× Dulbecco's phosphate-buffered saline (10× PBS).

7. Disposable polypropylene tubes (15 and 50 mL).

### 2.2. Culture of Rat Hepatocytes

1. Medium for spheroid formation: serum-free Williams' E medium supplemented with 10 mM nicotinamide, 10 ng/mL mouse EGF (Roche Diagnostics, Mannheim, Germany), and 10$^{-7}$ M insulin (Sigma Chemical Company, Saint Louis, MO).

2. Medium for three-dimensional collagen gel culture: Williams' E medium supplemented with nicotinamide, EGF, insulin, and 10% fetal bovine serum.

3. Collagen gel matrix: Cellmatrix type IA (Nitta gelatin).

4. Concentrated Williams' E medium: dissolve the powder of Williams' E medium (for 1 L) and 1.22 g of nicotinamide in 80 mL of distilled and deionized water. Make up to 100 mL with water. Do not add NaHCO$_3$.

5. Reconstruction buffer: 0.05N NaOH, 262 mM NaHCO$_3$, 20 mM Hepes.

6. Primaria dishes (Becton-Dickinson, Franklin Lakes, NJ) for the formation of spheroid aggregates.

7. Six-well or 12-well plastic plates for collagen gel culture.

## 3. Methods

### 3.1. Isolation of Rat Hepatocytes

1. Add 50 mg collagenase to 100 mL Solution II (final concentration: 0.05%), gently stir to dissolve completely, and filter through a 0.22-μm Corning filter (Collagenase solution). Warm Solution I (200 mL) and Collagenase solution to 40°C in a water bath.

2. Fix a deeply anesthetized rat on a cork board. After soaking the abdominal skin with 70% ethanol, open the abdomen, cannulate the portal vein, start perfusion with Solution I at a very low speed. Cut the vena cava inferior, increase the perfusion speed, and then perfuse with Collagenase solution (see Note 2).

3. After digestion, the entire liver is removed carefully and put into a plastic dish (10 cm in diameter). Add 20 mL Washing solution to the dish, tear the liver capsule with two pairs of fine tweezers, and disperse the liver tissue by gentle shaking in the solution. After pipetting up and down 20 times, add another 20 mL Washing solution. Then filter the diluted cell suspension through eightfold gauze to remove undigested debris.

4. Remove nonparenchymal cells by repeated low-speed centrifugations at $50–70 \times g$ (see Note 3). Resuspend the hepatocytic pellet with 15 mL of PBS, add 10 mL of 90% Percoll, and mix gently. After centrifugation at $150 \times g$ for 5 min, wash the pellet one time with PBS by low-speed centrifugation (see Note 4).

### 3.2. Formation of Hepatocytic Spheroids

1. Plate isolated hepatocytes onto Primaria dishes to form spheroidal aggregates (Fig. 2a) (see Note 5).

2. After 4 or 5 days, harvest spheroidal aggregates (Fig. 2b) by gentle pipetting or, if necessary, by scraping. Centrifuge at low speed and wash the pellet one time with PBS (see Note 6). Keep the tube containing spheroids on ice.

### 3.3. Collagen Gel Culture

1. Mix eight parts of Cellmatrix type I-A, one part of concentrated Williams' E medium, one part of the reconstruction buffer, and two parts of Williams' E medium (see Note 7). All the ingredients should be ice-cold and the neutralized collagen mixture should be kept on ice so as to avoid premature gelling (see Note 8).

2. Put the neutralized collagen mixture into the tube containing spheroids. Resuspend the spheroids by gentle pipetting. Place the tube on ice.

3. Carefully add an appropriate amount of the resuspended spheroids to the center of the well of 6-well or 12-well plastic plates and immediately spread it evenly by a pipet tip (see Note 9).

4. Put the plates with their lids in a $CO_2$ incubator at 37°C for 20 min for gelling.

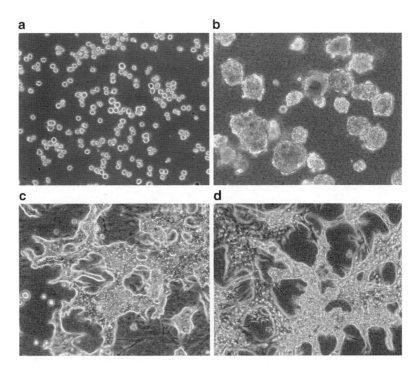

Fig. 2. Phase-contrast photographs of cultured hepatocytes. (**a**) Hepatocytes isolated from an adult rat liver by collagenase perfusion. (**b**) Hepatocytic spheroids formed on Primaria dishes. Immediately after being embedded within collagen gel. (**c**) Aggregates of hepatocytes showing branching morphogenesis within collagen gel after 14 days. The original spheroid shape is partially lost. (**d**) Extensive branching morphogenesis of hepatocytes treated with tumor necrosis factor (TNF)-α for 14 days. TNF-α is found to be a cytokine which enhances ductular transdifferentiation of hepatocytes.

5. Add serum-containing Williams' E medium to each well. Test compounds or factors, such as various cytokines, may be added at this point of time. Shake the plate to facilitate diffusion of the medium into the gel (see Note 10).

6. Start culturing and feed each culture with the medium every 2–3 days (Fig. 2c, d).

*3.4. Morphological Observation and Immunocytochemistry*

1. After completion of culturing, remove medium completely and wash each well once with PBS. Fix the cells within collagen gels with 10% buffered formalin for 12 h at room temperature (see Note 11).

2. Process paraffin-embedding, sectioning and staining with hematoxylin and eosin through standard procedures for tissues (Fig. 3).

3. For immunocytochemistry, deparaffinized sections are treated with an antigen-retrieval solution (such as Target Retrieval Solution [DAKO, Carpinteria, CA]). We perform immunocytochemistry by using LSAB kit for rat tissues (DAKO).

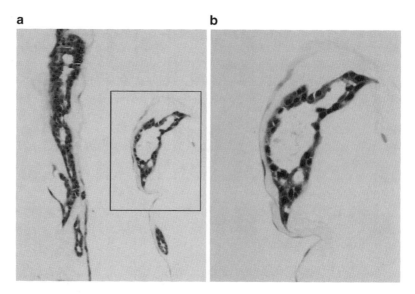

Fig. 3. Formation of ductular structures by hepatocytes within collagen gel for 14 days in the presence of TNF-$\alpha$. (**a**) Low magnification. (**b**) Higher magnification of the rectangular area in (**a**). There are irregularly shaped ductular structures, which are reminiscent of atypical ductular reaction in vivo.

4. For observation of the ultrastructures, cells should be fixed with 2.5% glutaraldehyde and 1% osmium tetroxide, and embedded in Epon resin. Ultrathin sections are stained with uranylate and lead, and observed under an electron microscope.

### 3.5. Western Blot Analysis

1. Wash gels containing cells with PBS and homogenize in a lysis buffer (such as RIPA buffer: 1% Triton X-100, 1% sodium deoxycholate, 0.1% SDS, 158 mM sodium chloride in 10 mM Tris–HCl buffer [pH 7.5)] containing protease inhibitors (see Note 12).

2. Measure protein concentration of each sample and prepare 40–50 μg aliquots.

3. Perform SDS-PAGE using 10% polyacrylamide gels and transfer the electrophoresed proteins to PVDF membranes.

4. Use standard Western blotting procedures for membrane blocking, incubation with primary and secondary antibodies. Detect signals with enhanced chemiluminescence reagents (Amersham, GE Healthcare).

### 3.6. Reverse Transcriptase-Polymerase Chain Reaction

1. Wash gels containing cells with PBS and homogenize in TRIzol reagent (Life Technologies, Invitrogen) or its equivalents (see Note 13).

2. Measure RNA concentration of each sample.

3. Perform reverse transcriptase-polymerase chain reaction (RT-PCR) analysis. We use a one-step RT-PCR kit (TaKaRa,

Ohtsu, Shiga, Japan), in which RNA (0.25–0.5 μg) was reverse transcribed using AMV reverse transcriptase for 30 min and then amplified for 25 cycles of 95, 60, and 72°C for 30, 30, and 90 s.

# 4. Notes

1. 100× Collagenase stock solution (5%, filtered) can be aliquotted and stored at −20°C for at least several months.

2. Perfusion with Solution I should be continued at least 10 min for efficient collagenase digestion.

3. We usually perform low-speed centrifugations three times in a 50-mL polypropylene tube.

4. Although this step can be omitted, it will significantly increase the viability of isolated hepatocytes.

5. We usually prepare cell suspension of $4.3 \times 10^5$ cells/mL and use 10 and 4 mL of cell suspension for a 10-cm dish and a 6-cm dish, respectively. Cell counting can be replaced with measurement of absorbance at 660 nm (at $4.3 \times 10^5$ cells/mL, $A_{660}$ will be 0.45).

6. Remove residual PBS as much as possible, so as not to reduce the final concentration of collagen gel matrix.

7. We first mix Cellmatrix A-1 and concentrated Williams' E medium, then add the reconstruction buffer and Williams' E medium. Addition of two parts of 1× Williams' E decreases the density of collagen gel matrix, which appears to facilitate branching morphogenesis. Mix thoroughly, but gently by stirring and pipetting with a disposable 1-mL pipet tip with the point ended being cut to avoid shearing collagen fibers.

8. Once collagen solution is neutralized by the reconstruction buffer, gelling takes place several minutes if they are left at room temperature. However, if collagen solution is placed on ice, gelling does not occur at least for several hours.

9. Spheroids obtained from one 10-cm Primaria dish are usually sufficient for three-dimensional culture on one plastic plate (6-well or 12-well). The volume of resuspended spheroids in collagen gel matrix for each well is 150 and 70 μL for 6-well and 12-well, respectively.

10. We usually add 2 and 1 mL of medium to each well of 6-well plates and 12-well plates, respectively.

11. Overfixation (e.g., more than 1 week) should be avoided since antigenicity of some antigens might be decreased or lost.

12. Since the medium contains serum proteins, gels should be vigorously washed with PBS. We wash gels in PBS by vortexing and collect them by centrifugation at 7,000–9,000×$g$ in a microcentrifuge. The washing should be repeated at least three times.

13. Washing with PBS can be brief or omitted for RNA extraction.

## References

1. Popper, H., Kent, G., Stein, R. (1957) Ductular cell reaction in the liver in hepatic injury. *J Mt Sinai Hosp NY* 24, 551–556.

2. Alvaro, D., Mancino, M. G., Glaser, S., Gaudio, E., Marzioni, M., Francis, H. et al (2007) Proliferating cholangiocytes: a neuroendocrine compartment in the diseased liver. *Gastroenterology* 132, 415–431.

3. Libbrecht, L., Desmet, V., Van Damme, B., Roskams, T. (2000) Deep intralobular extension of human hepatic 'progenitor cells' correlates with parenchymal inflammation in chronic viral hepatitis: can 'progenitor cells' migrate? *J Pathol* 192, 373–378.

4. Fausto, N., Campbell, J. S. (2003) The role of hepatocytes and oval cells in liver regeneration and repopulation. *Mech Dev* 120, 117–130.

5. Sell, S. (2001) Heterogeneity and plasticity of hepatocyte lineage cells. *Hepatology* 33, 738–750.

6. Van Eyken, P., Sciot, R., Desmet, V. J. (1989) A cytokeratin immunohistochemical study of cholestatic liver disease: evidence that hepatocytes can express 'bile duct-type' cytokeratins. *Histopathology* 15, 125–135.

7. Desmet, V., Roskams, T., Van Eyken, P. (1995) Ductular reaction in the liver. *Pathol Res Pract* 191, 513–524.

8. Shiojiri, N. (1997) Development and differentiation of bile ducts in the mammalian liver. *Microsc Res Tech* 39, 328–335.

9. Shiojiri, N., Nagai, Y. (1992) Preferential differentiation of the bile ducts along the portal vein in the development of mouse liver. *Anat Embryol (Berl)* 185, 17–24.

10. Nishikawa, Y., Tokusashi, Y., Kadohama, T., Nishimori, H., Ogawa, K. (1996) Hepatocytic cells form bile duct-like structures within a three-dimensional collagen gel matrix. *Exp Cell Res* 223, 357–371.

11. Nishikawa, Y., Doi, Y., Watanabe, H., Tokairin, T., Omori, Y., Su, M. et al (2005) Transdifferentiation of mature rat hepatocytes into bile duct-like cells in vitro. *Am J Pathol* 166, 1077–1088.

12. Michalopoulos, G. K., Bowen, W. C., Mule, K., Lopez-Talavera, J. C., Mars, W. (2002) Hepatocytes undergo phenotypic transformation to biliary epithelium in organoid cultures. *Hepatology* 36, 278–283.

13. Michalopoulos, G. K., Barua, L., Bowen, W. C. (2005) Transdifferentiation of rat hepatocytes into biliary cells after bile duct ligation and toxic biliary injury. *Hepatology* 41, 535–544.

# Part IV

**Liver Stem Cells and Hepatocarcinogenesis**

# Chapter 14

# Identification of Cancer Stem Cell-Related MicroRNAs in Hepatocellular Carcinoma

## Junfang Ji and Xin Wei Wang

## Abstract

Cancer Stem cells (CSCs) are the source of many solid tumor types including hepatocellular carcinoma. MicroRNAs (miRNAs) are small noncoding RNAs and have been showed to be associated with hepatic CSCs. Here, we described methods to screen hepatic CSC-related miRNAs, and to validate and examine their expressions and functions in vitro and in vivo, which contribute to the maintenance of stemness and differentiation of hepatic CSCs.

**Key words:** Hepatic cancer stem cells, MicroRNAs, Hepatocellular carcinoma, EpCAM, Fluorescence-activated cell sorting, MicroRNA microarray, Quantitative reverse transcription-polymerase chain reaction, Transfection, Spheroid assay, Tumorigenicity assay

## 1. Introduction

Cancer stem cells (CSCs) are cancer cells in tumor that possess characteristics associated with normal stem cells, i.e., self-renewal and differentiation, and the ability to give rise to a new tumor with the phenotype of original one in xenotransplant assays (1). They are identified as the source for cancer aggressive features in many solid tumor types including hepatocellular carcinoma (HCC) (2–7) so that targeting CSCs hold hope to eliminate cancer burden. HCC is the fifth most common and third most deadly malignancy worldwide with observable heterogeneity, and comprises about 90% of human liver cancers (8). Hepatic cancer stem cells (HpCSCs) have been characterized using a variety of stem cell markers, including epithelial cell adhesion molecule (EpCAM), CD90, and CD133 (7, 9, 10). HCC cases with high level of alpha fetal protein (AFP) and EpCAM have the poor prognosis and the progenitor cell gene

Takahiro Ochiya (ed.), *Liver Stem Cells: Methods and Protocols*, Methods in Molecular Biology, vol. 826,
DOI 10.1007/978-1-61779-468-1_14, © Springer Science+Business Media, LLC 2012

expression profiles, and EpCAM⁺ hepatic cancer cells isolated with an EpCAM-specific antibody by fluorescence-activated cell sorting (FACS) from AFP⁺ HCC cell lines or AFP⁺ HCC clinical specimens are HpCSCs. These cells display all three CSC characteristics by associated technologies, i.e., self-renewal by spheroid assay, differentiation by FACS assay, and initiating a tumor by tumorigenicity assay in NOD/SCID mice (7, 11).

MicroRNAs (miRNAs) are a novel class of small, noncoding RNAs that posttranscriptionally regulate gene expression through complementary base pairing to messenger RNAs (mRNAs) (12, 13). They are excellent biomarkers for cancer diagnosis, prognosis and therapy, and functionally involved in many biological processes including "stemness" (14–20). The miRNA pathway affects stem cell division and stem cell populations (21, 22). The expression of miRNA is developmentally regulated (23) and the distinct "stem cell" miRNA profiles are detected in pluripotent embryonic stem cells, normal stem cells, and CSCs from adult, but not in the normal differentiated cells (14, 24). Moreover, individual miRNAs have been shown to control cellular differentiation and self-renewal (25–27). In HCC, miR-181 family is found to be functionally associated with HpCSCs (14). It has also been found that therapeutic delivery of miRNAs suppresses HCC tumorigenesis in mice (28). Therefore, identifying liver CSC associated miRNAs may assist in targeting HpCSCs, which in turn may eradicate the tumor of HCC patients.

Many technologies have been developed to examine miRNA expression. Microarray technology is a powerful high-throughput tool to monitor the expression of thousands of miRNAs at once (29). Standard northern blotting is proposed for detecting and validating candidate miRNA, while it requires large amounts of total RNA (5–20 μg for each blot). Quantitative reverse transcription-polymerase chain reaction (qRT-PCR) is also used to examine miRNA expression, which is simple and robust, and only requires very small amounts of total RNA (10 ng for each RT reaction). Most recently, profiling of miRNAs by next generation sequencing technologies measures absolute abundance and allows for the discovery of novel miRNAs. And it can avoid previous cloning and standard sequencing efforts. Moreover, technologies have also been developed to explore the function of those small noncoding RNAs by altering their expression in vitro and in vivo, such as the vector-based miRNA expression systems (30) and antagomiR (31). A direct alteration of miRNA expression can be achieved by transfecting cells with miRNA precursor sequence (for miRNA overexpression) or the antisense of miRNA (for miRNA silencing).

Here, we described the methods of preparing cells for FACS to isolate EpCAM⁺ HpCSCs, miRNA microarray using isolated cells to screen HpCSCs-related miRNAs, qRT-PCR to examine and validate the expression of candidate miRNAs, transfection of oligos into cells to alter miRNAs' expression, spheroid assay, and tumorigenicity assay to explore their function. All the methods we outlined here were optimized in HuH7, a HCC cell line. The procedure of these experiments in other cells could be further modified according to the optimized condition in HuH7 cells and the emphasized notes in this chapter.

## 2. Materials

### 2.1. Prepare Cells for FACS

1. HBSS+ buffer (40 ml): Mix 38.8 ml Hank's balanced salt solution ($Ca^{2+}$ free, $Mg^{2+}$ free, and Phenol Red free), 0.8 ml FBS (2%), and 0.4 ml HEPES (1%) in 50 ml tube. Keep at 4°C for up to 1 month.

2. EpCAM antibody: FITC-anti-EpCAM (DAKO, Cat# F0860). Keep at 4°C from light.

3. 40-μm Cell strainer.

### 2.2. RNA Extraction

1. Trizol Reagent.

2. 75% Ethanol (40 ml): Mix 30 ml of 100% ethanol and 10 ml of DEPC-treated water in 50-ml tube. Keep at 4°C.

### 2.3. MicroRNA Microarray

1. HS4800 hybridization station (TECAN US).

2. Axon Scanner 4000B (Molecular Device).

3. Pre-hybridization mix (6× SSPE/2× Denhardts/30% Formamide) (5 ml): Add 1.5 ml of 20× SSPE, 200 μl of 50× Denhardt's solution and 1.5 ml of Formamide into 1.8 ml of deionized $H_2O$. Mix well. Prepare it before use.

4. 6× SSPE/30% Formamide (5 ml): Add 1.5 ml of 20× SSPE and 1.5 ml of formamide into 2.0 ml of deionized $H_2O$. Mix well. Prepare it before use.

5. 0.5 pmol/μl of (5′-biotin-$(dA)_{12}$-(dT-biotin)-$(dA)_{12}$-$(N)_8$-3′) custom random octomer oligonucleotide primer for reverse transcription (RT) reaction.

6. Superscript II RNase H⁻ reverse transcriptase (200 U/μl) with 5× first-strand buffer and 0.1 M dithiothreitol (DTT).

7. 1× TNT buffer (1,000 ml): Add 200 ml of 1 M Tris–HCl, 30 ml of 5 M NaCl, and 0.5 ml of Tween-20 into 769.5 ml of deionized $H_2O$. Mix well and filter it with a 0.2-μm filter. This solution can be stored up to 2 weeks at room temperature.

8. 0.75× TNT buffer (500 ml): Add 125 ml of deionized water to 375 ml of 1× TNT buffer (from above). Mix well and store up to 2 weeks at room temperature.

9. TNB buffer (500 ml): Mix 50 ml of 1 M Tris–HCl (pH 7.6), 15 ml of 5 M NaCl, and 2.5 g of NEN blocking reagent with 435 ml of nuclease-free H$_2$O. Incubate the mixture in a 60°C water bath to dissolve the blocking reagent. Filter TNB buffer through a 0.88-μm filter. Make 50 ml aliquots and store at −20°C for up to 12 weeks. Thaw it before use.

10. Streptavidin-Alexa 647 solution: Dissolve 1 mg of anhydrous Streptavidin-Alexa 647 conjugate in 1,000 μl of 1× PBS (pH 7.4) as stock solution. Store at 4°C in the dark for up to 2 weeks. For working solution, dilute the stock solution for 500 times in filtered TNB buffer. The working solution should be used within 15 min of preparation.

*2.4. qRT-PCR*

1. TaqMan MicroRNA Reverse Transcription Kit.
2. TaqMan microRNAs assays.
3. TaqMan Universal PCR Master Mix, No AmpErase UNG.

*2.5. Transfection*

1. Lipofectamine 2000.
2. Opti-MEM I Reduced-Serum Medium.
3. Oligos: Choose oligo for overexpressing the candidate miRNA from Pre-miR miRNA Precursors library and use the associated oligo as a negative control for experiments. Choose oligo for silencing the candidate miRNA from the library of Anti-miR miRNA inhibitors and use the associated oligo as the negative control. Briefly centrifuge the tube with oligos to ensure that the dried oligonucleotides are at the bottom of the tube. Resuspend the oligonucleotides with RNAse-free water to make the 50-μM stock solution. Aliquot and store at −20°C.

*2.6. Components for Examine miRNA's Function*

1. Ultra-Low Attachment six-well plate.
2. BD Matrigel Matrix High Concentration.
3. Kendall MONOJECT Hypodermic Needles (22 G × 1.5 in.).
4. Kendall MONOJECT Tuberculin Syringes without needle (1 ml).

# 3. Methods

*3.1. Prepare Cells for FACS*

1. Wash 85–95% confluent-cultured HuH7 cells vigorously with 1× PBS (pH 7.4) for two times.
2. Trypsinize cells with E-PET (1 ml/T75 flask) for 3 min at 37°C and make single cell suspension by pipetting cells in cold PBS. Add 1 ml of ice-cold culture media to stop trypsinization on ice, count cells by cell counter (see Note 1).

3. Take $2.0 \times 10^7$ cells into a 15-ml tube and $0.5 \times 10^6$ cells as control in the other tube (see Note 2).

4. Centrifuge by $200 \times g$ for 5 min at 4°C. Aspirate the supernatant. Wash cells with 500 μl of ice-cold HBSS+. Repeat one time of washing.

5. Suspend $2.0 \times 10^7$ cells with 500 μl of antibody solution (400 μl of HBSS+ and 100 μl of FITC-anti-EpCAM) and $0.5 \times 10^6$ in 500 μl of HBSS+ as negative control (see Note 3).

6. Incubate on ice for 30 min (protect from light). Gently mix cells every 10 min.

7. Centrifuge by $200 \times g$ for 5 min at 4°C. Aspirate the supernatant. Wash cells with 500 μl of ice-cold HBSS+ for two times.

8. Resuspend the stained cells with 1,000 μl of ice-cold HBSS+ and control cells with 500 μl ice-cold HBSS+. Filter cells through Cell strainer into tube (see Note 4).

9. Cell sorting in the FACS facility center (book an appointment in advance always).

10. Collect sorted cells by centrifuge with $200 \times g$ for 5 min at 4°C and wash by PBS for one time. Store cell pellets at −80°C until RNA isolation.

### 3.2. RNA Isolation (see Note 5)

1. Add Trizol to cell pellets from last step, and lyse cells by repetitive pipetting and followed vortex. Use 1 ml of the Trizol per $1–10 \times 10^6$ of isolated cells.

2. Incubate the homogenized samples for 5 min at room temperature.

3. Add 0.2 ml of chloroform per 1 ml of Trizol. Mix samples vigorously for 15 s until it appears as a uniform pink suspension. Incubate them at room temperature for 3 min (phase separation should be visible).

4. Centrifuge the samples at $10,000 \times g$ for 15 min at 4°C. Following centrifugation, the mixture separates into lower phenol–chloroform phase (Red), an interphase (White), and an upper aqueous phase (Colorless) with RNA.

5. Transfer upper aqueous phase into a fresh tube (see Note 6).

6. Add 0.5 ml of isopropyl alcohol per 1 ml of Trizol to the aqueous phase. Mix sample vigorously for 15 s, incubate samples at −20°C for 10 min, and centrifuge at $10,000 \times g$ for 10 min at 4°C (see Note 7).

7. Remove the supernatant completely. Wash the RNA pellet once with 75% ethanol (1 ml of 75% ethanol per 1 ml of Trizol). Mix the samples by gently upside down for five times, centrifuge at $6,000 \times g$ for 1 min at 4°C. Remove ethanol completely.

8. After air-drying or vacuum drying RNA pellet for 2 min, dissolve RNA in DEPC-treated water by pipetting.

9. Quantitate RNA by spectrophotometer (see Note 8). Store RNA samples at –80°C.

### 3.3. MicroRNA Microarray

#### 3.3.1. Prepare the Biotin-Labeled cDNAs for Hybridization

1. Take 5 μg of total RNA in 10 μl of RNase-free $H_2O$ and add 2 μl of 0.5 pmol/μl primer (5′-biotin-(dA)$_{12}$-(dT-biotin)-(dA)$_{12}$-(N)$_8$-3′) together in a total volume of 12 μl. Mix it by pipetting (see Note 9).

2. Incubate the mixture in 70°C water bath for 10 min. Place the reaction tube immediately on ice for 2 min. Briefly centrifuge reaction tube at 4°C to collect sample and keep it on ice till using (see Note 10).

3. Prepare the RT reaction mix on ice. The 8-μl per RT reaction is composed of 4 μl of 5× first-strand buffer, 2 μl of 0.1 M DTT, 1 μl of 10 mM dNTP mix, and 1 μl of Superscript II RNase H⁻ reverse transcriptase (200 U/μl). The reaction mixture should be scaled up proportionally on the basis of the number of samples to be handled for reverse transcription (see Note 11).

4. Add 8 μl of preprepared RT reaction mix to 12 μl of total RNA/primer mix. Mix the reaction mixture gently by pipetting up and down for several times and briefly spin the tube (see Note 12).

5. Incubate the reaction mix for 90 min in a 37°C water bath for synthesizing biotin-labeled first-strand cDNAs.

6. Centrifuge the tubes briefly at room temperature and put tubes onto ice.

7. Add 3.5 μl of 0.5 M NaOH/50 mM EDTA into the 20-μl RT reaction mix and incubate at 65°C for 15 min to denature the DNA/RNA hybrids and degrade single-strand RNA templates (see Note 13).

8. Add 5 μl of 1 M Tris–HCl (pH 7.6) to neutralize NaOH in reaction mix at room temperature. Store the labeled cDNAs at –20°C until use (see Note 14).

#### 3.3.2. MicroRNA Microarray

Hybridization of Biotin-labeled miRNA cDNAs with miRNA array slides (Ohio State University miRNA microarray version 4.0) was performed on Tecan HS4800 hybridization station (steps 1–6) under "microRNA expression" program.

1. Prime the chip in hybridization chamber at 23°C with 6× SSPE/0.5% Tween-20 for 1 min and soak for 1 min (see Note 15).

2. Inject 95 μl of prehybridization mix into hybridization chamber at 25°C. Prehybridize at 25°C for 30 min with medium agitation.

3. Inject 75 µl of hybridization mix (each labeled biotin-cDNA in 6× SSPE/30% formamide) into hybridization chamber. And hybridize for 18 h at 25°C with medium agitation.

4. Wash the array slides and chamber with 0.75× TNT buffer at 23°C for 5 min to remove the hybridization mix.

5. Wash the array slides with 0.75× TNT buffer at 37°C for a total of 9 min (1 min for two times followed by one time of 5 min and one time of 2 min) to remove unbound cDNAs on the slides.

6. Rinse with water at 23°C for 30 s to remove the salts of 0.75× TNT buffer.

7. Unload array slides from HS4800 and soak it in 37°C prewarmed 0.75× TNT buffer. Wash it in 37°C prewarmed 0.75× TNT buffer at 37°C for 40 min with agitation at $50 \times g$.

8. Block the array slides in 1× TNB blocking buffer at room temperature for 30 min.

9. Incubate slides with fresh prepared 1:500 Streptavidin-Alexa 647-TNB staining solution at room temperature for 30 min.

10. Wash with 1× TNT buffer at $50 \times g$. For total 40 min at room temperature in two fresh buffer changes.

11. Rinse array slides briefly with distilled water and transfer array slides onto metal slide rack. Spin-dry the array slide at $1,000 \times g$ for 1 min at room temperature.

12. Scan processed array slides with Axon 4000B scanner using red 635-nm laser at 10 µm resolution with Power 100 and PMT 800. And export the miRNA reading.

## 3.4. qRT-PCR

After analyzing the microarray data from last step, choose the candidate liver CSCs-associated miRNAs for further validation with qRT-PCR.

### 3.4.1. RT (Multiplex Reaction)

1. Dilute 1 µg total RNAs to 10 ng/µl with nuclease-free water. Place on ice (see Note 16).

2. Prepare the RT master mix on ice according to the desired number of RT reactions and add 15% overage to account for pipetting losses. The 11-µl per RT reaction is composed of 8.16 µl of nuclease-free water, 0.15 µl of 100 mM dNTPs, 1 µl of 50 U/µl MultiScribe Reverse Transcriptase, 1.5 µl of 10× Reverse Transcription Buffer, and 0.19 µl of 20 U/µl RNase Inhibitor. Mix gently but completely by pipetting. Place on ice.

3. Mix RT master mix from last step with total RNA in the ratio of 11:1 by gently pipetting upside down (see Note 17).

4. Transfer 12 µl of the mixture of step 3 in a 96-well-plate well; add 3 µl of 5× RT primer from each TaqMan MicroRNA Assay set into the corresponding well.

5. Mix gently by tapping and centrifuge $200 \times g$ at 4°C for 3 min. Place it on ice for 5 min (see Note 18).

6. Synthesize the cDNA in a thermocycler according to following temperatures: 16°C for 30 min, 42°C for 30 min, and 85°C for 5 min.

7. Store the cDNA products at −20°C until use.

*3.4.2. qPCR*

1. Prepare the qPCR reaction mix according to the desired number of PCR reactions. Add 10% overage to account for pipetting losses. The 18.67-µl per qPCR reaction mix is composed of 7.67 µl of nuclease-free water, 10 µl of 2× TaqMan Universal PCR Master Mix, 1 µl of 20× Real Time Primer from each TaqMan MicroRNA Assay (see Note 19).

2. Vortex the mixture and briefly centrifuge. Transfer 18.67 µl of the mixture to a 96-well-plate well.

3. Add 1.33 µl of RT products from each RT reaction into the corresponding well.

4. Perform PCR reaction according to the following cycling program in a real-time PCR machine: 95°C for 10 min, 40 cycles of 95°C for 10 s, and 60°C for 1 min.

**3.5. Transfection**

1. The day before transfection, plate 235,000 HuH7 cells to a six-well plate well in 2 ml complete fresh medium according to 25,000 cells/cm² so that they will be 70–75% confluent on the day of transfection (see Note 20).

2. On the day of transfection, remove the culture medium from the cells and replace with 1 ml of fresh medium (without antibiotics) for each well (see Note 21).

3. For each transfection, prepare oligo-Lipofectamine 2000 complexes as follows:

   (a) Prepare oligo solution in Opti-MEM. For one well of HuH7 cells in six-well plate, mix 30 µl of Opti-MEM with 30 pmol oligos gently by tapping and then incubate at room temperature for 5 min (see Note 22).

   (b) Prepare Lipofectamine 2000 solution in Opti-MEM. For one well of HuH7 cells in six-well plate, mix 30 µl of Opti-MEM with 1.5 µl of lipofectamine 2000 gently by tapping and then incubate this solution at room temperature for 5 min.

   (c) Combine the oligo solution with the prepared Lipofectamine solution. Mix gently by tapping. Incubate for 20 min at room temperature to allow the oligo Lipofectamine complexes to form. The solution may appear cloudy.

4. Add the oligo–Lipofectamine complexes to each well. Mix gently by rocking the plate back and forth. Incubate the cells for 6 h at 37°C in a humidified 5% $CO_2$ incubator.

5. After 6 h, remove the medium containing the oligo–Lipofectamine complexes and replace with 2 ml complete culture medium.

6. Two days later, collect cells for examining miRNA expression with qRT-PCR or for the following experiments.

*3.6. Spheroid Assay*

1. Collect the cells with altering miR expression from the last step by trypsinization. Make single cell suspension and count cells.

2. Seed 1,000 cells in each well of Ultra-Low Attachment six-well plate with 2 ml complete medium. Mix gently by rocking the plate back and forth.

3. Culture for 10–12 days at 37°C in a humidified incubator with 5% $CO_2$.

4. Count spheroids under microscope.

*3.7. Tumorigenicity Assay*

1. Two days before injection, isolate EpCAM[+] cells according to the methods we described above. Seed isolated EpCAM[+] HpCSCs (235,000 cells/well) in six-well plate for overnight.

2. Transfect EpCAM[+] HpCSCs with the oligos (precursors or antimiRs) according to the methods part (Subheading 3.5).

3. Thaw Matrigel on ice in 4°C freezer overnight and freeze all 1 ml syringes and 22 G needles in –20°C (see Note 23).

4. One day later of transfection, collect cells by trypsinization. Make single cell suspension and count cells.

5. For ten-site injection, transfer $2.0 \times 10^4$ cells to a 15-ml tube. Centrifuge at $200 \times g$ for 5 min at 4°C. Dispose all the supernatant and suspend them in 2 ml precooled culture medium. Keep cells on ice.

6. Aspirate 2 ml Matrigel to a syringe without attaching a needle. Mix Matrigel and cell suspension well quickly on ice (final concentration as 1,000 cells/200 μl) (see Note 24).

7. Keep all cell–Matrigel mixture, 1 ml syringes and 22 G needles on ice until use. Inject cell–Matrigel mixture into the subcutaneous tissues of NOD/SCID mice (see Note 25).

# 4. Notes

1. Cells must be in a single-cell suspension. For adherent cells, over-trypsinization or less-trypsinization could cause cell clumps, which could be removed by passing through a cell-strainer.

2. We expect one million sorted EpCAM⁺ HpCSCs and EpCAM⁻ differentiated cells, which will allow us to obtain enough RNA (more than 5 μg). The top 5% EpCAM⁺ and top 5% EpCAM⁻ cells will be collected. Therefore, $2.0 \times 10^7$ cells will be needed for cell sorting.

3. For staining of certain amount of cells, calculate the reagent and the buffer following 500 μl of HBSS+ buffer including 100 μl of antibodies per $2.0 \times 10^7$ cells. Keep the prepared antibody solution from light and avoid over-exposure to the light during the procedure of steps 7–11.

4. Resuspend cells routinely at an approximate concentration of $1.0 \times 10^7$ cells/ml for the actual sort. The highest density of cells is $2.0 \times 10^7$ cells/ml. For cells that tend to clump excessively, a cell concentration of approximately $0.5 \times 10^7$ can reduce clumping.

5. General considerations when working with RNA: wear gloves at all times; use RNase-free tips and tubes for all samples (autoclaved and not touched without gloves); use RNase-free solutions.

6. The volume of the aqueous phase is about 60% of the volume of Trizol used for homogenization. We would expect to get 50% of the volume of Trizol. Be careful NOT to transfer material from the interface layer – it is better to lose a little RNA than to risk contamination of the whole sample.

7. Low temperature is helpful to facilitate RNA precipitation. If the cell number is very low, RNA could be precipitated at this step for overnight. The RNA precipitate, often invisible before centrifugation, forms a gel-like pellet on the side and bottom of the tube.

8. The ratio of absorbance at 260 and 280 nm (A260/A280) is used to assess the purity and quantity of RNA. RNA with a ratio of ~2.0 is generally accepted as "pure" for RNA. A ratio of ~1.8 is considered to be contaminated by DNA. A ratio of <1.8 is considered to be contaminated by protein. A ratio of >2.1 is considered to be degraded. The ratio of absorbance at 260 and 230 nm (A260/A230) is used as a secondary measure of nucleic acid purity. A ratio of 1.9–2.2 is considered as "Good" for RNA. A ratio of <1.8 is considered to be contaminated by phenol. Moreover, total RNAs with a 28S:18S rRNA ratio of 2.0 or greater are also considered as good quantity. However, it is rare to see. Generally, total RNA with 28S:18S rRNA ratio >1.0 and a low baseline between the 18S and 5S rRNA is also considered as good RNA. If RNA quality is considered to be "bad" based on the assessment above, it is strongly suggested to repeat the cell sorting and RNA isolation.

9. In the standard operating procedure, 5 µg of total RNA is needed for detecting miRNAs expressed at a low level.

10. Reaction takes place at 70°C for high efficiency and specificity, without secondary structure on RNA templates. It is necessary to immediately transfer the tube on ice from 70°C water bath and keep the tube on ice until it cools down.

11. Make reaction mixes fresh on ice and always add the reverse transcriptase as the last step. To minimize variability of the RT reaction, it is important that RT reaction mixtures for all samples are prepared together instead of individually.

12. Avoid air bubbles in this mixture. Air bubbles in the reaction mix can affect the efficiency of RT reaction.

13. Degrading the RNA templates completely is important to avoid competitive hybridization to biotin-labeled miRNA cDNAs with oligo probes on the array.

14. The labeled biotin-cDNA can be stored at −20°C for months until use.

15. Avoid miRNAs array slides (or chips) being scratched. Avoid slides being attached by air bubbles by always Tap slide rack. Avoid slide getting exposed to air, as slide drying affects image quality due to increased background noise.

16. Do not denature the RNA. Denaturation of the RNA may reduce the yield of cDNA for some miRNA targets.

17. Take the RT mixture (11 µl for each reaction) according to the desired number of RT-primers with 10% overage into a 1.5-ml tube. Add RNA (10 ng/µl, 1 µl for each reaction) according to the desired number of RT-primers with 10% overage.

18. Do not centrifuge over $500 \times g$ or 5 min.

19. Keep all 20× Real time primers of TaqMan MicroRNA Assays protected from light in the freezer until using. Excessive exposure to light may affect the fluorescent probes. Mix the TaqMan Universal PCR Master Mix thoroughly by swirling the bottle prior to use. Moreover, all the qPCR reactions need to be triplicating at least.

20. Too much or less cells are not good. For other cells, optimized condition is needed to make sure that certain cells will be 70–75% confluent on the day of transfection.

21. Reducing the amount of cell culture medium in this step could increase the transfection efficiency. A half of regular adding amount is suggested.

22. RNA oligonucleotides are susceptible to degradation by exogenous ribonucleases introduced during handling. Wear gloves when handling this product. Use RNase-free reagents, tubes, and barrier pipette tips. Upon receipt, store at or below −20°C.

To minimize freeze–thaw cycles, prepare a concentrated 50 µM stock, then further dilute to a practical working stock concentration.

23. Since Matrigel matrix forms a gel above 10°C, this solution should be kept at low temperature. And all reagents and equipments being contacted with Matrigel should be chilled on ice prior to injection. For injection, enough syringes and needles (three for each cell) should be prepared and a needle with proper size should be selected to prevent the destruction of cells.

24. Add 100 µl of Matrigel per plug to 100 µl of cell suspension and mix gently to avoid foaming. For each plug, 200 µl of cell–Matrigel mixture is needed. Prepare the mixture according to the desired plugs with 100% overage. Therefore, 4 ml of cell–Matrigel mixture should be prepared for ten plugs.

25. To increase the contact area of the injected Matrigel mixture into subcutaneous tissues, a wide subcutaneous pocket should be formed by swaying the needlepoint right and left after a routine subcutaneous insertion. Then the Matrigel mixture was injected into this pocket. When the BD Matrigel mixture is injected into a particular area without swaying the needle point, the mixture will form a large cell clump and a subsequent growth defect may result due to inefficient perfusion of nutrients to the cells within the core of the clump.

## Acknowledgment

This work was supported by the Intramural Research Program of the Center for Cancer Research, the National Cancer Institute (Z01 BC 010313 and Z01 BC 010876).

## References

1. Gupta PB, Chaffer CL, and Weinberg RA (2009) Cancer stem cells:mirage or reality? Nat. Med 15:1010–1012

2. Al Hajj M, Wicha MS, Benito-Hernandez A et al (2003) Prospective identification of tumorigenic breast cancer cells. Proc. Natl. Acad. Sci USA 100:3983–3988

3. Singh SK, Hawkins C, Clarke ID et al (2004) Identification of human brain tumour initiating cells. Nature 432:396–401

4. Bonnet D. and Dick JE (1997) Human acute myeloid leukemia is organized as a hierarchy that originates from a primitive hematopoietic cell. Nat. Med. 3:730–737

5. Ricci-Vitiani L, Lombardi DG, Pilozzi E et al (2007) Identification and expansion of human colon-cancer-initiating cells. Nature 445:111–115

6. O'Brien CA, Pollett A, Gallinger S et al (2007) A human colon cancer cell capable of initiating tumour growth in immunodeficient mice. Nature 445:106–110

7. Yamashita T, Ji J, Budhu A et al (2009) EpCAM-positive hepatocellular carcinoma cells are tumor-initiating cells with stem/progenitor cell features. Gastroenterology 136:1012–1024

8. Parkin DM, Bray F, Ferlay J et al (2005) Global cancer statistics, 2002. CA Cancer J. Clin 55:74–108

9. Yang ZF, Ho DW, Ng MN et al (2008) Significance of CD90(+) Cancer Stem Cells in Human Liver Cancer. Cancer Cell 13:153–166

10. Ma S, Chan KW, Hu L et al (2007) Identification and characterization of tumorigenic liver cancer stem/progenitor cells. Gastroenterology 132:2542–2556

11. Yamashita T, Forgues M, Wang W et al (2008) EpCAM and alpha-fetoprotein expression defines novel prognostic subtypes of hepatocellular carcinoma. Cancer Res. 68:1451–1461

12. Lagos-Quintana M, Rauhut R, Lendeckel W et al (2001) Identification of novel genes coding for small expressed RNAs. Science 294:853–858

13. Lau NC, Lim LP, Weinstein EG et al (2001) An abundant class of tiny RNAs with probable regulatory roles in Caenorhabditis elegans. Science 294:858–862

14. Ji J, Yamashita T, Budhu A et al (2009) Identification of microRNA-181 by genome-wide screening as a critical player in EpCAM-positive hepatic cancer stem cells. Hepatology 50:472–480

15. Ji J, Shi J, Budhu A et al (2009) MicroRNA expression, survival, and response to interferon in liver cancer. N. Engl. J Med 361:1437–1447

16. Ji J. and Wang X.W (2009) New kids on the block:Diagnostic and prognostic microRNAs in hepatocellular carcinoma. Cancer Biol Ther. 8:1686–1693

17. Budhu A, Jia HL, Forgues M et al (2008) Identification of metastasis-related microRNAs in hepatocellular carcinoma. Hepatology 47:897–907

18. Lu J, Getz G, Miska EA et al (2005) MicroRNA expression profiles classify human cancers. Nature 435:834–838

19. Croce CM. and Calin GA (2005) miRNAs, cancer, and stem cell division. Cell 122:6–7

20. Rogler CE, Levoci L, Ader T et al (2009) MicroRNA-23b cluster microRNAs regulate transforming growth factor-beta/bone morphogenetic protein signaling and liver stem cell differentiation by targeting Smads. Hepatology 50:575–584

21. Hatfield SD, Shcherbata HR, Fischer KA et al (2005) Stem cell division is regulated by the microRNA pathway. Nature 435:974–978

22. Bernstein E, Kim SY, Carmell MA et al (2003) Dicer is essential for mouse development. Nat. Genet. 35:215–217

23. Strauss WM, Chen C, Lee CT et al (2006) Nonrestrictive developmental regulation of microRNA gene expression. Mamm. Genome 17:833–840

24. Suh MR, Lee Y, Kim JY et al (2004) Human embryonic stem cells express a unique set of microRNAs. Dev. Biol. 270:488–498

25. Chen CZ, Li L, Lodish HF et al (2004) MicroRNAs modulate hematopoietic lineage differentiation. Science 303:83–86

26. Schratt GM, Tuebing F, Nigh EA et al (2006) A brain-specific microRNA regulates dendritic spine development. Nature 439:283–289

27. Yi R, Poy MN, Stoffel M et al (2008) A skin microRNA promotes differentiation by repressing 'stemness'. Nature 452:225–229

28. Kota J, Chivukula RR, O'donnell KA et al (2009) Therapeutic microRNA delivery suppresses tumorigenesis in a murine liver cancer model. Cell 137:1005–1017

29. Liu CG, Calin GA, Meloon B et al (2004) An oligonucleotide microchip for genome-wide microRNA profiling in human and mouse tissues. Proc Natl Acad Sci USA 101:9740–9744

30. Voorhoeve PM, le Sage C, Schrier M et al (2006) A genetic screen implicates miRNA-372 and miRNA-373 as oncogenes in testicular germ cell tumors. Cell 124:1169–1181

31. Krutzfeldt J, Rajewsky N, Braich R et al (2005) Silencing of microRNAs in vivo with 'antagomirs'. Nature 438:685–689

# Part V

## Application of Liver Stem Cells for Cell Therapy

# Chapter 15

# Intravenous Human Mesenchymal Stem Cells Transplantation in NOD/SCID Mice Preserve Liver Integrity of Irradiation Damage

## Moubarak Mouiseddine, Sabine François, Maâmar Souidi, and Alain Chapel

## Abstract

This work was initiated in an effort to evaluate the potential therapeutic contribution of the infusion of mesenchymal stem cells (MSC) for the correction of liver injuries. We subjected NOD–SCID mice to a 10.5-Gy abdominal irradiation and we tested the biological and histological markers of liver injury in the absence and after infusion of expanded human MSC. Irradiation alone induced a significant elevation of the ALT and AST. Apoptosis in the endothelial layer of vessels was observed. When MSC were infused in mice, a significant decrease of transaminases was measured, and a total disappearance of apoptotic cells. MSC were not found in liver. To explain the protection of liver without MSC engraftment, we hypothesize an indirect action of MSC on the liver via the intestinal tract. Pelvic or total body irradiation induces intestinal absorption defects leading to an alteration of the enterohepatic recirculation of bile acids. This alteration induces an increase in Deoxy Cholic Acid (DCA) which is hepatotoxic. In this study, we confirm these results. DCA concentration increased approximately twofold after irradiation but stayed to the baseline level after MSC injection. We propose from our observations that, following irradiation, MSC infusion indirectly corrected liver dysfunction by preventing gut damage. This explanation would be consistent with the absence of MSC engraftment in liver. These results evidenced that MSC treatment of a target organ may have an effect on distant tissues. This observation comes in support to the interest for the use of MSC for cellular therapy in multiple pathologies proposed in the recent years.

**Key words:** Human, Mesenchymal stem cells, Liver, Irradiation exposure, NOD/SCID mice, PCR, AST, ALT, Liver bilic acids, Therapeutic transplantation

## 1. Introduction

Mesenchymal stem cells (MSC) are described as multipotent progenitor cells that differentiate into osteocytes, chondrocytes, adipocytes, and stromal cells (1–3). MSC have been successfully used

Takahiro Ochiya (ed.), *Liver Stem Cells: Methods and Protocols*, Methods in Molecular Biology, vol. 826,
DOI 10.1007/978-1-61779-468-1_15, © Springer Science+Business Media, LLC 2012

in therapeutics (4, 5) to correct osteogenesis imperfecta (6), to improve haematopoiesis (7–9) and to prevent Graft Versus Host Disease post-haematopoietic stem cell transplantation (10–12).

In addition to these therapeutical domains, MSC have recently received attention for their potential as regenerative medicine following radiation injuries. The ileum of irradiated pigs shows many alterations such as blunting and/or villi loss. Radiation-induced alterations of intestinal absorption have been previously described (13, 14).

In the rat model, it has been shown that radiation induces acute alterations of the enterohepatic recirculation which concomitantly with radiation-induced intestinal malabsorption, leads to alterations of hepatic synthesis and secretion (15). In NOD–SCID mice, intestinal radiation injury is characterized by impaired epithelial renewal, leading to mucosal disruption and functional abnormalities (16). We have previously described the capacity of MSC to restore intestinal integrity after radiation-induced damage (17). We have also shown that MSC when infused to mice that received either extended or localized irradiation, migrate to almost all tissues where they engraft transiently usually at very low levels of detection (18). Observations in the liver of human transplant recipients have shown a long-term implantation of a very small number of donor bone marrow cells which might be of mesenchymal origin (19). MSC improve to liver function and modulate hepatocellular death. These effects may be mediated by hepatocytes replacement and/or secretion of growth factors (20, 21).

In this work, we found that human MSC, when injected after abdominal irradiation in NOD–SCID mice, were able to indirectly preserve the liver of radiation damage. We studied the mechanism of this protection. Following our previous findings on intestinal protection, we found that the MSC regenerated the small intestine epithelium, which in turn restored the enterohepatic recirculation pathway initially damaged by irradiation. The consequence was a distant hepatic protection without engraftment of MSC in liver.

## 2. Materials and Methods

### 2.1. Culture of Human MSC

Bone marrow (BM) cells were obtained from iliac crest aspirates of healthy volunteers after informed consent and were used in accordance with the procedures approved by the human experimentation and ethic committees of the *Hôpital St Antoine*. As previously described (22), 50 ml of BM were taken from different donors in the presence of heparin (Sanofi-synthélabo, France). Low-density mononuclear cells (MNC) were separated on the Ficoll Hypaque density gradient (d 1.077). MNC were plated at a density of $1.33 \times 10^6$ cells/cm$^2$ corresponding to a concentration of

$10^7$ cells/10 ml of McCoy's 5A medium supplemented with 12.5% foetal calf serum, 12.5% horse serum, 1% sodium bicarbonate, 1% sodium pyruvate, 0.4% MEM non-essential amino acids, 0.8% MEM essential amino acids, 1% MEM vitamin solution, 1% l-glutamine (200 mM), 1% penicillin–streptomycin solution (all from Invitrogen, Groningen, The Netherlands), $10^{-6}$ M hydrocortisone (Stem Cell Technologies®), 2 ng/ml human basic recombinant fibroblast growth factor (R&D System, Abington, UK) in T-75 cm² tissue culture flasks and incubated at 37°C in humidified, 5% $CO_2$ atmosphere. After 3 days, non-adherent cells were washed with PBS and fresh medium (without hydrocortisone) was added. Samples of human MSC from different donors were collected at the second passage for transplantation.

### 2.2. FACS Analysis

Before injection, the MSC phenotype was checked. Stainings were performed with phycoerythrin (PE)-conjugated monoclonal antibody against CD105 (SH2), CD73 (SH3), and CD45 (Becton, Dickinson and Company, Franklin Lakes, NJ, http://www.bd.com) for 30 min at 4°C followed by two washes in PBS containing 0.5% BSA. Cells were re-suspended in 200 μl of PBS, 0.5% BSA, and analyzed at 10,000 events per test by FACScalibur BD (Pharmingen, San Diego, CA, USA). Mouse immunoglobulin G1 (IgG1) was used as an isotopic control (Beckman Coulter, Fullerton, CA, USA).

### 2.3. Configuration of Irradiation

Mice were irradiated locally at the abdominal area using an ICO4000 device (Cobalt 60 source). The window of irradiation is 2 cm of width and 3 cm of length. The dose rate is 2 Gy/min. After anaesthesia mice were maintained on target during the time of irradiation. Each mouse was separately irradiated at the dose of 10.5 Gy.

### 2.4. NOD/SCID Mice Model

All experiments and procedures were performed in accordance with the French Ministry of Agriculture regulations for animal experimentation (Act n°87-847 October 19, 1987, modified May 2001) and were approved by the animal care committee of the Institut de Radioprotection et de Sûreté Nucléaire (IRSN). NOD-LtSz*scid/scid* (NOD–SCID) mice, from breeding pairs originally purchased from Jackson Laboratory (Bar Harbor, ME, USA) were bred in our pathogen-free unit and maintained in sterile microisolator cages. A total of 25 12-week-old mice divided into four groups were used for this study. Two groups received IV a dose of $5 \times 10^6$ human MSC. Group 1 was a nonirradiated control group and did not receive human MSC. Group 2 was not irradiated and received MSC infusion. Group 3 was irradiated at a sublethal dose of 10.5 Gy and did not receive MSC infusion. Group 4 was irradiated at a sublethal dose of 10.5 Gy and received MSC infusion 5 h later.

### 2.5. Quantitative PCR

The animals were sacrificed 5 days after irradiation. Peripheral blood, liver, and kidneys were collected, and the quantitative implantation of human MSC in mouse tissue was defined by real-time PCR experiments as previously described (23). Briefly human beta globin gene was amplified from DNA in order to detect human cell engraftment; mouse RAPSYN gene was used as control of amplification.

### 2.6. Histology

Liver and kidney alterations were studied by histological analysis on day 5.

Formalin-fixed, paraffin-embedded liver and kidney from NOD/SCID mice were cut at 5 μm on a rotary microtome (Leica Microsystems AG, Wetzlar, Germany) and mounted on polysine slides. Sections were deparaffinized in xylene and rehydrated with ethanol and PBS. Sections were stained with haematoxylin, eosin, and safran (HES).

### 2.7. Tunnel Assay

Apoptotic cells were determined using the in situ cell Death Detection kit from Roche Diagnostics (Mannheim, Germany) following the manufacturer's instructions. The apoptotic cells (brown staining) were counted under a microscope. The apoptotic index was defined by the percentage of brown (dark) cells among the total number of cells in each sample. Five fields with 100 cells per field were randomly counted for each sample. We counted a minimum of three samples, thus making a total of 15 single analyses.

### 2.8. Plasma Analysis

Animals were anaesthetized by Rompan-Imalgene solution and killed by intracardiac puncture with a 1-ml insulin syringe to collect blood. The abdomen was opened and the liver and kidney were rapidly excised, weighed, and apportioned for preparing cellular fractions or storage at −80°C for future use. Blood was collected in 1.5-ml tubes with heparin syringe to prevent serum formation by in vitro coagulation. Samples were centrifuged at $6{,}000 \times g$ at 4°C for 5 min to collect the plasma. We used an automated Konelab 20 apparatus (Thermo Electron Corporation, Courtaboeuf, France) system to measure plasma alanine amino-transferase (ALT), aspartate amino-transferase (AST) (biological chemistry reagents, Bayer Diagnostics) in the control and different groups of mice.

### 2.9. Quantification of Liver Bilic Acids

Liver samples were previously homogenized in 150 mM NaCl. The procedure of sample extraction is adapted from Keller et al. Norcholic acid was used as the internal standard. Analysis of bilic acids (BA) was performed by gas chromatography–mass spectrometry (GC-MS).

### 2.10. Statistical Analysis

All values were expressed as the mean and SEM (standard error of the mean). To compare results between groups, we used a T-test or a one-way ANOVA followed by Tukey test with Sigmastat software (Systat Software Incorporation). Significance for all analyses was set at ***$p < 0.001$ or **$p < 0.05$.

# 3. Results

### 3.1. MSC Characterization

Phenotypic analysis showed that the MSC used in these experiments following expansion, were strongly positive for the specific surface antigens SH2 and SH3, respectively 37.3% ± 4.0 and 72.9% ± 3.7. Almost no contamination (0.2% ± 0.1 CD 45+ cells) by haematopoietic cells was detected (Table 1). MSC had their specific fibroblast-like appearance. Before use, we checked that each batch of MSC retained its specific ability to undergo osteogenic, chondrogenic, and adipogenic differentiation. Our results suggest that the MSC used in these experiments for transplant were expanded without significant loss in their differentiation capacities.

### 3.2. MSC Were Not Found in Liver

Human MSC were tracked in liver and gut at 5 days after injection using real-time PCR. MSC engrafted in small intestine (0.2% ± 0.03) but not in liver.

### 3.3. MSC Have No Hepatotoxic Potential

There was no significant variation in ALT (108 ± 18.9 U/I) and AST (20.3 ± 0.8 U/I) levels when we compared control mice with mice receiving MSC (ALT = 113 ± 43.5 U/I, AST = 17.4 ± 1.7 U/I) (Fig. 1a, b).

## Table 1
**Sample analysis and phenotypic characterization of cultured-expanded human MSC: Phenotypic analysis of MSC was performed at the second passage, before transplant**

| No.° | BM (ml) | MSC ($10^6$ cell) (s) | % CD73 | % CD105 | % CD45 |
|------|---------|----------------------|--------|---------|--------|
| 1    | 25.0    | 37.5                 | 65.9   | 38.1    | 0.3    |
| 2    | 20.0    | 79.0                 | 72.5   | 43.1    | 0.3    |
| 3    | 8.5     | 20.5                 | 71.6   | 48.6    | 0.4    |
| 4    | 30.0    | 19.4                 | 67.7   | 27.7    | 0.1    |
| 5    | 32.0    | 58.5                 | 87.0   | 28.9    | 0.1    |
| Mean | 23.1    | 43.1                 | 72.9   | 37.3    | 0.2    |
| SEM  | 4.2     | 11.5                 | 3.7    | 4.0     | 0.1    |

Frequency of positive cells for specific markers of hMSC: SH2 (CD105) and SH3 (CD73) and haematopoietic cell markers (CD45)

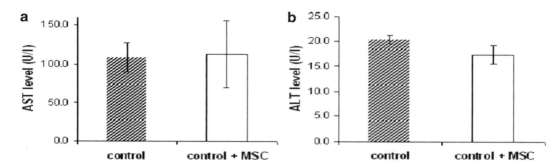

Fig. 1. Absence of toxicity of human MSC infused to NOD/SCID mice: Hatched histograms correspond to control animals non-irradiated receiving no MSC ($n=7$). *White* histograms correspond to non-irradiated animals receiving MSC ($n=4$). There was no difference between the two groups for plasma levels of transaminases AST (**a**) and ALT (**b**).

### 3.4. Irradiation Has a Hepatotoxic Effect

Abdominal irradiation resulted in a significant elevation of the ALT ($353.9 \pm 37.5$ vs. $108.6 \pm 18.9$ U/I) and AST ($40.3 \pm 4.9$ vs. $20.3 \pm 0.8$ U/I) levels after 5 days ($p < 0.001$) (Fig. 2a, b). Therefore, hepatic function was altered 5 days after abdominal irradiation. By TUNEL marking on the liver before and after irradiation, we observed that irradiation induced apoptosis of the endothelial cells and the cells lining the bile ducts which are polarized hepatocytes. Irradiation induced damage to the liver (Fig. 3a, b).

### 3.5. MSC Protect the Liver After Irradiation

MSC infusion prevented AST ($198.8 \pm 43.1$ U/I) and ALT ($14.9 \pm 4.7$ U/I) from increasing 5 days after irradiation ($p < 0.05$ and $p < 0.01$, respectively). (Fig. 2a, b).

### 3.6. MSC Injection Limits Liver Apoptosis (Fig. 3c)

Five days after irradiation, apoptosis was investigated in the liver. In the untreated group, irradiation-induced apoptosis in the endothelial layer of vessels was observed ($29\% \pm 7$ of apoptotic cells). However, when mice were transplanted with MSC the apoptotic areas disappeared ($3\% \pm 2$ apoptotic cells, $p < 0.001$ compared to untreated group, Table 2).

### 3.7. MSC Protect the Enterohepatic Recirculation Pathway

Intestinal malabsorption is associated with an alteration of the enterohepatic recirculation of bile acids. In Fig. 4, we only present the variation of the most hydrophobic acid, DCA, in so far as the others did not vary significantly. After abdominal irradiation, bile acid concentrations were disturbed. The results show that DCA concentration increased approximately twofold. In contrast, after MSC injection the concentration of DCA equalled the control level. MSC therefore can prevent a 5-day increase in deoxycholic acid concentration. MSC can correct the radio-induced disturbances which cause the variation of DCA. Since bile acids are regulated by the bowel, MSC can restore intestinal absorption of DCA.

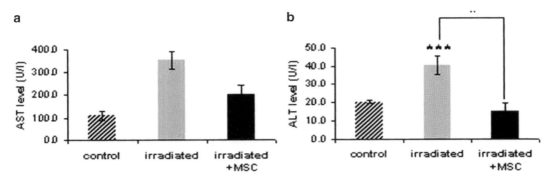

Fig. 2. *Effect of irradiation and MSC infusion on liver.* Hatched histograms correspond to non-irradiated control animals receiving no MSC ($n = 7$). *Grey histograms* correspond to irradiated animals (dose: 10.5 Gy) with no MSC infusion ($n = 5$). *Black histograms* correspond to animals receiving MSC after irradiation ($n = 8$). Plasma dosages were done on day 5 post-irradiation. (**a**) Transaminases AST: a significant increase in the AST plasma concentration was observed which was multiplied by threefold ($p < 0.001$). MSC infusion resulted in a significant reduction of the increase ($p < 0.05$). (**b**) Transaminases ALT: a significant increase in the ALT plasma concentration was observed which was multiplied by twofold ($p < 0.01$). MSC infusion resulted in a significant reduction of the increase ($p < 0.05$).

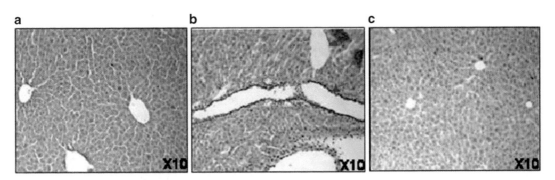

Fig. 3. *Evaluation of apoptosis in the liver 5 days after irradiation followed or not by MSC infusion.* TUNEL staining: apoptotic and necrotic cells appear brown-coloured (*red arrows*). (**a**) Control, the necrotic and apoptotic cells seen after irradiation alone (**b**) were not seen in animals receiving MSC (**c**) post-irradiation (magnification ×10).

## Table 2
## Percentage of apoptotic cells 5 days after irradiation in liver

| Percentage of apoptotic cells | Irradiated | | | Irradiated + MSC | | |
|---|---|---|---|---|---|---|
| 5 days after irradiation | Mean | SEM | *n* | Mean | SEM | *n* |
| Liver | 29 | 7 | 5 | 3 | 2 | 8 |

In the untreated group (irradiated), irradiation-induced apoptosis. In mice transplanted with MSC (irradiated + MSC) the apoptotic cells disappeared ($p < 0.001$ compared to untreated group)

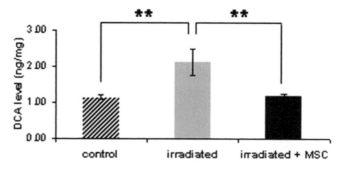

Fig. 4. *Deoxycholic acid (DCA) variation 5 days after irradiation followed or not by MSC infusion*: Hatched histograms correspond to non-irradiated control animals (*n* = 7), *grey histograms* to irradiated animals, *black histograms* to animals receiving MSC after irradiation (*n* = 11). DCA concentration in liver increased significantly ($p < 0.01$), by a factor of 2, 5 days after irradiation. After infusion of MSC, DCA concentration did not increase. The significant result for irradiated versus irradiated and treated is $p < 0.01$.

## 4. Discussion

This work was initiated in an effort to evaluate the potential therapeutic contribution of the infusion of MSC for the correction of liver injuries induced by an irradiation exposure.

We subjected NOD–SCID mice to a 10.5-Gy abdominal irradiation and we tested the biological and histological markers of liver injury in the absence and after infusion of expanded human MSC. Irradiation alone induced a significant elevation of the ALT and AST. Necroses endothelial cells and the cells lining the bile ducts were observed. Infusion of MSC into animals 5 h after irradiation induced a significant decrease of transaminases, and an almost total disappearance of apoptotic cells. An effect of MSC with a decrease of apoptosis has, however, been described in a different situation, namely acute renal failure (ARF) (24).

To explain the restoration of liver damage as evaluated by the correction of transaminase levels and the decreasing number of apoptotic cells in the liver, as well as the absence of detectable implantation of MSC in the liver at 5 days after infusion, we hypothesize an indirect action on the liver via the intestinal tract. Total body irradiation and abdominal irradiation induce structural and functional gut damage (16, 24, 25) with a decrease of villi height (17) and malabsorption (15). Intestinal malabsorption in turn induces an alteration of the enterohepatic recirculation of bile acids (15), with an increase in Deoxy Cholic Acid (DCA) which is described as one of the most cytotoxic of them (22). Indeed, in this study, DCA concentration increased approximately twofold after irradiation but went back to the baseline level after MSC.

We propose that, following irradiation, MSC infusion indirectly corrected liver dysfunction by preventing gut damage. One mechanism for this indirect effect is the alteration of the enterohepatic recirculation pathway. Another mechanism that should be considered is the role of cytokines and growth factors produced by the MSC that are homing to other organs. Recent publications by Van Poll et al. showed in a model of fulminant hepatic failure (FHF) suggest that the infusion of conditioned medium from MSC protects from apoptosis in the liver and stimulate hepatocyte proliferation (20). The indirect effect of MSC on the liver reported in this study could correspond to a similar mechanism. This explanation would be consistent with the absence of detection of MSC in the liver. We believe our observation brings an additional piece of evidence in favour of the use of MSC in patients submitted to pelvic or total body irradiation.

## Author Disclosure

No competing financial interests exist.

## References

1. Friedenstein AJ, Gorskaja JF, Kulagin NN. et al (1976) Fibroblast precursors in normal and irradiated mouse hematopoietic organs. Exp Hematol 4:267–274

2. Prockop DJ. (1997). Marrow stromal cells as stem cells for nonhematopoietic tissues. Science 276:71–74

3. Pittenger MF, Mackay AM, Beck SC, et al (1999) Multilineage potential of adult human mesenchymal stem cells. Science; 284:143–147

4. Lazarus HM (1995) Bone marrow transplantation in low-grade non-Hodgkin's lymphoma. Leuk Lymphoma 17:199–210

5. Koc ON, Peters C, Aubourg P, et al (1999) Bone marrow-derived mesenchymal stem cells remain host-derived despite successful hematopoietic engraftment after allogeneic transplantation in patients with lysosomal and peroxisomal storage diseases. Exp Hematol 27:1675–1681

6. Horwitz EM, Gordon PL, Koo WK, et al (2002) Isolated allogeneic bone marrow-derived mesenchymal cells engraft and stimulate growth in children with osteogenesis imperfecta: Implications for cell therapy of bone. Proc Natl Acad Sci USA. 99:8932–8937

7. Koc ON, Gerson SL, Cooper BW, et al (2000) Rapid hematopoietic recovery after coinfusion of autologous-blood stem cells and culture-expanded marrow mesenchymal stem cells in advanced breast cancer patients receiving high-dose chemotherapy. J Clin Oncol. 18:307–316

8. Fouillard L, Bensidhoum M, Bories D, et al (2003) Engraftment of allogeneic mesenchymal stem cells in the bone marrow of a patient with severe idiopathic aplastic anemia improves stroma. Leukemia. 17:474–476

9. Fouillard L, Chapel A, Bories D, et al (2007) Infusion of allogeneic-related HLA mismatched mesenchymal stem cells for the treatment of incomplete engraftment following autologous haematopoietic stem cell transplantation. Leukemia 21:568–570

10. Lazarus HM, Loberiza FR., Zhang MJ, et al (2001) Autotransplants for Hodgkin's disease in first relapse or second remission: a report from the autologous blood and marrow transplant registry (ABMTR). Bone Marrow Transplant 27:387–396

11. Le Blanc K, Rasmusson I, Sundberg B, et al (2004) Treatment of severe acute graft-versus-host disease with third party haploidentical mesenchymal stem cells. Lancet 363:1439–1441

12. Ringden O, Uzunel M, Rasmusson I, et al. (2006) Mesenchymal stem cells for treatment

of therapy-resistant graft-versus-host disease. Transplantation 81:1390–1397

13. Thomson AB, Cheeseman CI, Walker K et al (1984). Effect of external abdominal irradiation on the dimensions and characteristics of the barriers to passive transport in the rat intestine. Lipids 19:405–418

14. Yeoh E, Horowitz M, Russo A, et al (1994) Effect of pelvic irradiation on gastrointestinal function: a prospective longitudinal study. Am J Med 95:397–406

15. Scanff P, Grison S, Marais T, et al (2002) Dose dependence effects of ionizing radiation on bile acid metabolism in the rat. Int J Radiat Biol 78:41–47

16. Francois S, Bensidhoum M, Mouiseddine M, et al (2006) Local irradiation not only induces homing of human mesenchymal stem cells at exposed sites but promotes their widespread engraftment to multiple organs: a study of their quantitative distribution after irradiation damage. Stem Cells 24:1020–1029

17. Semont A, Francois S, Mouiseddine M, et al (2006) Mesenchymal stem cells increase self-renewal of small intestinal epithelium and accelerate structural recovery after radiation injury. Adv Exp Med Biol 585:19–30

18. Bensidhoum M, Chapel A, Francois S, et al (2004) Homing of *in vitro* expanded Stro-1- or Stro-1+ human mesenchymal stem cells into the NOD/SCID mouse and their role in sup-

porting human CD34 cell engraftment. Blood 103:3313–3319

19. Korbling M, Katz RL, Khanna A, et al (2002) Hepatocytes and epithelial cells of donor origin in recipients of peripheral-blood stem cells. N Engl J Med 346:738–746

20. Poll DV, Parekkadan B, Cho CH, et al (2008) Stem cell-derived molecules directly modulate hepathocellular death and regeneration in vitro and in vivo. Hepathology 47: 1634–1643

21. Togel F, Hu Z, Weiss K, et al (2005) Administered mesenchymal stem cells protect against ischemic acute renal failure through differentiation-independent mechanisms. Am J Physiol Renal Physiol 289:F31-42

22. Hofmann AF (1999) The continuing importance of bile acids in liver and intestinal disease. Arch Intern Med 159:2647–2658

23. Banas A, Teratani T, Yamamoto Y, et al (2008) IFATS collection: in vivo therapeutic potential of uman adipose tissue mesenchymal stem cells after transplantation into mice with liver injury. Stem cells 26: 2705–2712

24. Quastler H. (1956) The nature of intestinal radiation death. Radiat Res 4:303–320

25. Francois A, Milliat F, Vozenin-Brotons MC, et al (2003) 'In-field' and 'out-of-field' functional impairment during subacute and chronic phases of experimental radiation enteropathy in the rat. Int J Radiat Biol 79:437–450

# Chapter 16

# Engineering of Implantable Liver Tissues

**Yasuyuki Sakai, M. Nishikawa, F. Evenou, M. Hamon, H. Huang, K.P. Montagne, N. Kojima, T. Fujii, and T. Niino**

## Abstract

In this chapter, from the engineering point of view, we introduce the results from our group and related research on three typical configurations of engineered liver tissues; cell sheet-based tissues, sheet-like macroporous scaffold-based tissues, and tissues based on special scaffolds that comprise a flow channel network. The former two do not necessitate in vitro prevascularization and are thus promising in actual human clinical trials for liver diseases that can be recovered by relatively smaller tissue mass. The third approach can implant a much larger mass but is still not yet feasible. In all cases, oxygen supply is the key engineering factor. For the first configuration, direct oxygen supply using an oxygen-permeable polydimethylsiloxane membrane enables various liver cells to exhibit distinct behaviors, complete double layers of mature hepatocytes and fibroblasts, spontaneous thick tissue formation of hepatocarcinoma cells and fetal hepatocytes. Actual oxygen concentration at the cell level can be strictly controlled in this culture system. Using this property, we found that initially low then subsequently high oxygen concentrations were favorable to growth and maturation of fetal cells. For the second configuration, combination of poly-l-lactic acid 3D scaffolds and appropriate growth factor cocktails provides a suitable microenvironment for the maturation of cells in vitro but the cell growth is limited to a certain distance from the inner surfaces of the macropores. However, implantation to the mesentery leaves of animals allows the cells again to proliferate and pack the remaining spaces of the macroporous structure, suggesting the high feasibility of 3D culture of hepatocyte progenitors for liver tissue-based therapies. For the third configuration, we proposed a design criterion concerning the dimensions of flow channels based on oxygen diffusion and consumption around the channel. Due to the current limitation in the resolution of 3D microfabrication processes, final cell densities were less than one-tenth of those of in vivo liver tissues; cells preferentially grew along the surfaces of the channels and this fact suggested the necessity of improved 3D fabrication technologies with higher resolution. In any case, suitable oxygen supply, meeting the cellular demand at physiological concentrations, was the most important factor that should be considered in engineering liver tissues. This enables cells to utilize aerobic respiration that produces almost 20 times more ATP from the same glucose consumption than anaerobic respiration (glycolysis). This also allows the cells to exhibit their maximum reorganization capability that cannot be observed in conventional anaerobic conditions.

**Key words:** 3D culture, Cellular sheet, Macroporous scaffold, Flow channel, Oxygen, Respiration

Takahiro Ochiya (ed.), *Liver Stem Cells: Methods and Protocols*, Methods in Molecular Biology, vol. 826,
DOI 10.1007/978-1-61779-468-1_16, © Springer Science+Business Media, LLC 2012

## 1. Introduction

Relatively simple hepatocyte-based therapies have successfully been applied to human clinical trials. For instance, bioartificial livers can carry out liver functions at least in the short term and can be used as a bridge to liver transplantation (1). Simple hepatocyte infusion has been applied to human patients with metabolic diseases such as Criger–Najjar syndrome (2). However, long-term and full substitution of overall liver functions have not been achieved so far. Therefore, liver tissue equivalents based on tissue engineering methodologies are eagerly expected as recently reviewed by Ohashi et al. (3). Such preorganization of hepatocytes in vitro should overcome the very low initial engraftment of cells simply infused as suspensions (4).

There are three major types of engineered liver tissues in terms of their culture configuration; cell-sheet-based tissues without any scaffold (Fig. 1a), tissues based on sheet-like macroporous or fibrous biodegradable polymer scaffolds (Fig. 1b), and tissues based on scaffolds comprising a flow channel network as an artificial vasculature (Fig. 1c). This classification is largely decided by how oxygen and nutrients are supplied to the engineered tissues. In the sheet-like configurations (Fig. 1a, b), oxygen/nutrients

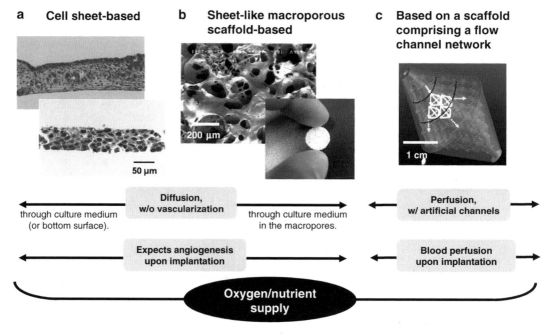

Fig. 1. Three typical engineered liver tissues for implantation. (**a**) Multilayered cell sheet; (**b**) 3D culture using thin macroporous biodegradable polymer scaffolds; (**c**) 3D culture using macroporous biodegradable scaffolds comprising a flow channel network.

should be supplied by simple diffusion processes and this limits the thickness of the sheet-like engineered tissues. This thickness can easily be predicted from the oxygen diffusion and consumption in high cell density tissues (5), and in vivo it is well known that almost all the cells reside within 200 μm from the nearest blood vessels (6). In contrast, in the special scaffold that includes a flow channel network (Fig. 1c), where culture medium or blood are continuously perfused, this limitation is removed, allowing the organization of much larger tissue masses.

Due to the technical or biological difficulties, the former two configurations (Fig. 1a, b) are widely used for implantation purposes because they do not necessitate tissue prevascularization, which has not yet been achieved to date. Instead, angiogenesis from the surrounding tissues is expected to occur quickly upon in vivo implantation. The most representative approach of the first category (Fig. 1a) is cell-sheet engineering based on tissue production and its noninvasive recovery using thermoresponsive polymer-immobilized plates. This enables the recovery of intact cell sheets from the plate surface by lowering the temperature without conventional enzymatic digestion that simultaneously breaks up cell-to-cell contacts and accumulated ECM in the cell sheet (3). Thick 3D liver tissues have also been prepared this way by repeated stacking of 2D hepatocyte sheets that could be transplanted and were viable until in vivo vascular formation occurred (7). Mooney and coworkers have produced pioneering work in the second category (Fig. 1b) (8). Recently, their initial scaffold was improved to simultaneously release three growth factors, vascular endothelial growth factor (VEGF), epidermal growth factor (EGF), and hepatocyte growth factor (HGF), but use of adult hepatocytes showed some limitations (9). The third configuration (Fig. 1c) inherently presupposes the connection of some artificially arranged vasculature in the scaffold to the host vasculature upon implantation, in almost the same manner as during organ transplantation. This enables immediate replacement of the functions in the hosts, but still requires future extensive studies concerning the arrangement of complete and stable endothelialization of the engineered tissues (10).

In any case, oxygen supply is the most determining factor, from the engineering point of view, in deciding the overall cell density in all these configurations. In this chapter, we introduce and summarize the results from our group and related research. First, we describe materials (Subheading 2) and methods for cell culture (Subheading 3) and notes for specific experimental points that should be paid attention to in order to obtain robust biological data (Subheading 4). Second, regarding the first category (Fig. 1a), we specifically describe the advantages of oxygen-permeable materials such as polydimethylsiloxane (PDMS), a kind of transparent silicone elastomer, as a culture surface to obtain thick liver tissues in conventional 2D plate cultures (Subheading 5). Regarding the second category (Fig. 1b), we point out the importance of the uses of progenitor hepatocytes to obtain high cell density in the scaffolds (Subheading 6). Concerning the third category (Fig. 1c), we introduce the design concept of artificial flow channel networks based on oxygen supply, 3D fabrication of the scaffold and cell culture using the scaffold (Subheading 7) for the next generation of engineered liver tissues. Finally, we present a general conclusion (Subheading 8) and expected future works (Subheading 9).

## 2. Materials

### 2.1. Established Cell Lines

Human hepatocarcinoma Hep G2 cells and NIH 3T3 cells were obtained from JCRB (Japanese Cancer Research Bank).

### 2.2. Animals

Adult and pregnant rats were all purchased from Sankyo Labo Service, Tokyo, Japan. The use of adult and pregnant rats was approved by the Animal Experimentation Committee at the Institute of Industrial Science, University of Tokyo, and the experiments were conducted according to the University of Tokyo guidelines for animal experimentation. Porcine fetuses were obtained from a local slaughterhouse (Tokyo Meat Market, Tokyo, Japan) through an agency supplying animal tissues for research (Tokyo Shibaura Internal-Organs Inc., Tokyo, Japan).

### 2.3. Enzymes for Isolation of Primary-Cultured Hepatocytes

For the isolation of adult rat hepatocytes, preperfusion and digestion media containing collagenase are prepared according to the protocol of Seglen (11). For the isolation of fetal hepatocytes, preperfusion buffer containing EGTA and digestion medium containing collagenase and dispase are all purchased from Invitrogen (Carlsbad, CA, USA). Tris(hydroxymethyl)aminomethane chloride (Tris–Cl) and $NH_4Cl$ for the hypotonic hemolysis buffer were purchased from Wako (Osaka, Japan).

### 2.4. Culture Media, Growth Factors, and Other Important Supplements

All basal culture media (William's E and DMEM) and phosphate-buffered saline (PBS) were purchased from Invitrogen (Carlsbad, CA, USA). Hydroxyethylpiperazine-$N'$-2-ethanesulfonic acid (HEPES) was purchased from (Dojindo, Kumamoto, Japan). Fetal bovine serum (FBS) was purchased from Gemini Bio-Product (West Sacramento, CA, USA). Mouse epidermal growth factor (mEGF), fibroblast growth factors (FGF) 1 and 4, and HGF were purchased from Peprotech (Rocky Hill, NJ, USA). Mouse Oncostatin M (OSM) was purchased from R&D systems (Minneapolis, MN, USA). Minimum essential medium (MEM) nonessential amino acid (NEAA) solution and antibiotic/antimycotic solution were purchased from Invitrogen. Glucagon and hydrocortisone were purchased from Sigma (St. Louis, MO, USA). l-Glutamine, insulin, dexamethasone, nicotinamide, sodium butyrate ascorbic acid 2-phosphate, and other trace elements for serum-free culture were purchased from Wako (Osaka, Japan).

**2.5. Tissue Culture Ware and Scaffold Materials**

Usual polystyrene tissue culture ware was purchased from Sumitomo Bakelite (Osaka, Japan). Type-I collagen for culture surface coating was purchased from Nitta Gelatin (Osaka, Japan). PDMS and its curing agent (Silpot 184) were purchased from Dow Corning (Tokyo, Japan). The photo-reactive cross-linker, Pierce SAND (P-C21549) was purchased from Thermo Fisher Scientific Inc. (Rockford, IL, USA). Poly-l-lactic acid (PLLA) (M.W. = 300,000) for 3D macroporous scaffolds was purchased from Polyscience (Warrington, PA, USA). Its solvent chloroform and $NH_4HCO_3$, the porogen for the formation of PLLA scaffolds, were purchased from Wako (Osaka, Japan). Poly-ε-caprolactone (PCL) (M.W. = 50,000) powders were a gift from Daicel Chemical Industry Ltd. (Osaka, Japan).

**2.6. Reagents for Biochemical or Histological Analyses**

Nonlabeled and peroxidase-labeled antibodies for rat and porcine albumin, and albumin standard for each species were purchased from Cappel Product (Aurora, OH, USA). Hematoxylin and eosin solutions were purchased from Wako (Osaka, Japan).

# 3. Methods for Cell Isolation and Culture

## 3.1. Culture of Hep G2 and NIH 3T3 Cells

1. Hep G2 and NIH 3T3 cells were routinely cultured in DMEM with a low glucose content culture medium containing 10% FBS, 1% NEAA, and an antibiotic/antimycotic solution. For Hep G2 cell culture on PDMS membranes (Subheading 5) and perfusion culture in a special 3D scaffold (Subheading 6), DMEM with a high glucose content additionally supplemented with 0.5 mM ascorbic acid 2-phosphate was used (see Note 2).

2. For coculture with adult rat hepatocytes, NIH 3T3 cells were resuspended with the culture medium employed for adult rat hepatocyte culture (Subheading 3) after supplementation with 5% FBS and inoculated at a density of $1.0 \times 10^5$ cells/cm$^2$ on tissue culture-treated polystyrene (TCPS) or PDMS membranes in a 12-well plate format at 8 h before or 24 h after hepatocyte inoculation (12).

## 3.2. Isolation and Culture of Adult Rat Hepatocytes

1. Adult rat hepatocytes were isolated from Wistar male rats weighing 200–250 g by the conventional two-step collagenase perfusion digestion method by Seglen (11).

2. For their culture, serum-free DMEM was basically used except during cocultivation with NIH 3T3 cells. The DMEM was supplemented with 1% NEAA, 10 ng/mL mEGF, 0.1 µM insulin, 0.1 µM dexamethasone, 0.8 µM copper sulfate ($CuSO_4 \cdot 5H_2O$), 2 nM selenium acid ($H_2SeO_3$), 2.6 µM zinc sulfate ($ZnSO_4 \cdot 7H_2O$), 0.3 µM manganese sulfate ($MnSO_4 \cdot 5H_2O$), 0.5 mM ascorbic acid 2-phosphate, and an antibiotic/antimycotic solution.

3. Hepatocytes were inoculated on collagen-precoated TCPS plates (12- or 24-well) at a density of $1.1 \times 10^5$ cells/cm$^2$. For the coculture with 3T3 cells, 5% FBS was added to the serum-free DMEM (12).

### 3.3. Isolation and Culture of Fetal Rat Hepatocytes (13)

1. After ether anesthesia of E17 pregnant Wistar rats, fetuses were removed and collected in a 90-mm dish with PBS. Usually 8–12 fetuses were obtained from one rat. Livers were then removed from each fetus and placed in a new dish.

2. The livers were minced by surgical scissors and washed with liver perfusion medium at 38°C for 10 min. After centrifuging them once at $60 \times g$ for 3 min, the minced liver tissues were treated with liver digest medium for 15 min in a water bath at 38°C. Liver cells were dispersed by gentle pipetting. RBCs were disrupted with a hypotonic hemolysis buffer (1 g Tris–Cl and 2.8 g NH$_4$Cl were dissolved in 500 mL water and sterilized). Finally, after filtration with a 70-µm Falcon cell strainer, hepatocytes were obtained by two centrifugations at $60 \times g$ for 1 min.

3. Hepatocytes were cultured in Williams' medium E supplemented with 2 mM l-glutamine, $10^{-6}$ M hydrocortisone, 10 ng/mL mEGF, $10^{-7}$ M insulin, $10^{-8}$ M glucagon, 0.5 mM ascorbic acid 2-phosphate, antibiotic/antimycotic solution, 1% MEM NEAA solution, 10% FBS, 10 mM nicotinamide, 10 ng/mL OSM, 20 ng/mL FGF-1, 20 ng/mL FGF-4, 20 ng/mL HGF, and 1 mM sodium butyrate. They were then inoculated at a density of $5–8 \times 10^4$ cells/cm$^2$ on TCPS or PDMS membranes in a 12- or 24-well plate format (14) or $5 \times 10^5$ cells/PLLA disc in a six-well plate on a rotational shaker (see Notes 1, 2 and 3).

### 3.4. Isolation and Culture of Fetal Porcine Hepatocytes (15)

1. Porcine fetuses (700–1,200 g) were stored in PBS on ice and transported to the laboratory within 4 h after they were removed from their mother.

2. Hepatocytes were isolated by a two-step perfusion technique as previously described by Seglen (11). With the vena cava and portal vein ligated, the liver was perfused via the umbilical vein as follows: first with preperfusion medium at a flow rate of less than 30 mL/min for 10 min, second with 0.05% collagenase-containing buffer for 5 min at 38°C. The digested liver tissues

were further incubated in the collagenase digestion buffer with gentle agitation for an additional 10 min. Then, the partly digested tissues were gently pipetted and mechanically dispersed in serum and growth factor-free MEM to produce a cell suspension. The suspension was centrifuged at 800 rpm for four times and filtered through a 70-µm Falcon cell strainer.

3. The cells were cultured in Williams' medium E supplemented with 2 mM l-glutamine, $10^{-6}$ M hydrocortisone or dexamethasone, 10 ng/mL mEGF, $10^{-6}$ or $10^{-7}$ M insulin, antibiotic/antimycotic solution, and 10% FBS. Hep G2 cells were cultured with high glucose DMEM supplemented with 10% FBS, 1% NEAA, 0.5 mM ascorbic acid 2-phosphate, and an antibiotic/antimycotic solution. The cells were inoculated at $4.0 \times 10^5$ cells per piece in a 0.03% collagen-coated PLLA scaffold in bacterial grade polystyrene 6-well plates and at $5.0 \times 10^4$ cells/$cm^2$ in collagen-precoated 12-well plates. The PLLA culture was continuously shaken using a rotational shaker at a rotational speed of 60 rpm (see Notes 1 and 3).

**3.5. Biochemical Measurements of Cellular Functions**

Albumin secreted into the culture medium was measured with the sandwich-type enzyme-linked immunosorbent assay (ELISA) using a horseradish peroxidase-conjugated secondary antibody. Glucose concentrations in culture medium were measured with a glucose analyzer (GA05, A&T Corp., Japan). Lactate concentrations were measured using the YSI 7100 Multiparameter Bioanalytical System analyzer (YSI Incorporated, Yellow Springs, OH).

**3.6. Histological Observation**

Cell-loaded PDMS membranes or 3D scaffolds were fixed in paraformaldehyde solution and embedded in paraffin for cross-section analysis with hematoxylin and eosin (HE).

# 4. Notes

1. The most crucial issue in ensuring successful start-up of culture is cell counting in the final suspension after isolation/purification steps. Unlike adult hepatocytes, the size of fetal hepatocytes is very small and not distinguishable from nonparenchymal cells. We disrupt RBCs using a hypotonic hemolysis buffer, but there still remain some RBCs and WBCs. Therefore, we need to strictly count fetal hepatocytes having intracellular granules or organelles, while neglecting nonparenchymal cells that have less or almost no granules or organelles, and RBCs and WBCs that are round and have no granules or organelles. However, manual counting of hepatic cells depends on the experimenter. Therefore, the best practical way is to establish a counting criterion through actual monolayer culture on collagen-coated dishes using an appropriate culture

medium. In our case, $5 \times 10^4$ cells/cm² is routinely employed as the minimum inoculation density for consistently starting up successful cultures of all species including mice (16), rats (13), and pigs (15). Inoculation at $2 \times 10^4$ cells/cm² sometimes resulted in failure, that is, instead of hepatic cells, nonparenchymal cells appeared dominant after several days of culture.

2.     All widely known gas-permeable polymers have a high hydrophobic property, which is not readily suitable for usual cell culture, because this property enables less adsorption of cell attachment proteins. We routinely do plasma treatments that disrupt the natural polymer structure and add some functional groups such as –OH or –COOH. Collagen adsorption by a usual coating procedure is enhanced enough to allow the polymer to serve as a suitable cell culture surfaces for Hep G2 cells or adult hepatocytes in coculture with NIH 3T3 cells (Subheading 5). However, this simple surface treatment is not enough in pure culture of adult rat hepatocytes or fetal rat hepatocytes; those cells gradually form aggregates, become less adherent to the surface and finally float in the culture medium, with once-adsorbed collagen molecules in the aggregates. This results in the loss of highly functional cells from the culture during usual culture medium replenishments. Covalent binding of collagen molecules can inhibit such aggregation/floating; adult rat hepatocytes form stably attached hemispheroids (17) and fetal rat hepatocytes can form thick attached cell layers (14) as shown later in Fig. 8 and described in Subheading 5.

Another issue in multilayered liver-derived cell culture is the nutrient supply. Oxygen-permeable PDMS membrane-based culture accommodates significantly higher numbers of cells when compared with usual TCPS culture. Therefore, we need to pay attention to the consumption of important nutrients such as glucose. For Hep G2 multilayered culture, we used DMEM with a high glucose content (4.5 g/L) and did daily medium replenishment. In the case of fetal rat hepatocytes, glucose consumption was not a problem in William's medium E (2.2 g/L) with a medium exchange on every other day. The selection of culture medium in terms of glucose content and feeding schedule should be determined by the remaining glucose concentrations in the spent culture medium.

3.     The first issue is the inoculum cell number to such 3D scaffolds. In the case of the small-scale 3D culture using a collagen-coated PLLA disc (10 mm in diameter, 1.2 mm in thickness; 0.10 cm³ in volume), as to be described in detail in Subheading 6, we just put one disc and cell suspension (2 mL) in six-well plate and immediately start rotational shaking (13). This inoculum usually gives an almost 100% attachability of the inoculated cells after 24 h of culture. Usually the minimum volumetric inoculation density is at

least $10^6$ cells/cm$^3$ scaffold. Another method is to keep the same surface cell density in 3D scaffolds as that in monolayers. Namely, we can roughly evaluate the total inner surface area of the PLLA disc scaffold from the mean diameter, specific density, and amount of $NH_4HCO_3$ particles used as a porogen. In the case of the PLLA disc (0.1 cm$^3$), this total inner surface area is about 10 cm$^2$. Therefore, $5.0 \times 10^4$ cells/cm$^2$, which is the standard inoculation density of fetal rat hepatocyte cultures, is converted to $5 \times 10^5$ cells/disc or $5 \times 10^6$ cells/cm$^3$ scaffold. In the case of perfusion culture using special scaffolds (Subheading 7), the inoculum density should be enhanced to ensure successful start-up of the culture. In our case using Hep G2 cells (18) or fetal porcine hepatocytes, we employed at least $10^7$ cells/cm$^3$ scaffold. In addition, simple rotational culture cannot be used for such a large perfusable scaffold. One good way to seed the cells is to inject a cell suspension, close and apply centrifugation forces to make cells enter into the deeper sites of the scaffolds (18), whereas simple repeated perfusion of cell suspensions cannot get cells inside the scaffolds.

Oxygen and nutrient supply should also be optimized for continuous stable culture. For glucose, we need to check at least the remaining concentration in the spent culture medium and decide the culture medium in terms of glucose content and frequency of medium exchange. Oxygen supply is more crucial but rather difficult to control. In the case of shaking culture of small disk-shaped scaffolds, cellular growth is still limited around the inner surface of macropores of the scaffold, suggesting insufficient oxygen supply to the deeper sites (Subheading 6). In the case of perfusion culture, we need to pay attention to both the oxygen concentration at the outlet of the scaffold and the flow rates that determine the shear stress in the scaffold, as discussed later in Subheading 7.

## 5. Engineering of Thick Liver Tissues Using an Oxygen-Permeable Membrane in Static Culture

### 5.1. Oxygen Supply in Static Culture

A study in the 1960s pointed out the fact that a confluent hepatocyte monolayer cultured in TCPS is usually put under an extremely anaerobic condition (19). This can easily be predicted from Fick's First law, simple oxygen diffusion through the culture medium layers and the oxygen consumption of the hepatocyte monolayer beneath the culture medium as shown in Fig. 2. This is mainly caused by almost ten times higher oxygen consumption by hepatocytes compared with that of fibroblasts (20). As a result, oxygen concentration at the cell layer is calculated to be almost 0 mol $O_2$/cm$^3$.

Uses of oxygen permeable membranes are the simplest method to overcome this limitation. Perfluorocarbon-based membranes partly succeeded in improving the in vitro functions of hepatocytes

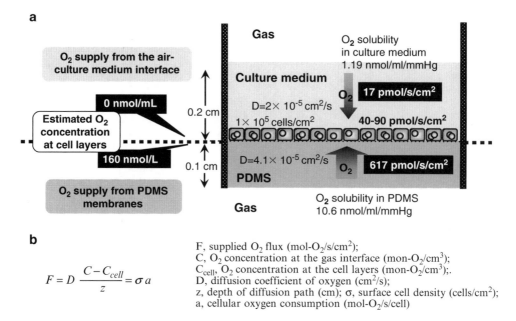

**b**

$$F = D \, \frac{C - C_{cell}}{z} = \sigma a$$

F, supplied $O_2$ flux (mol-$O_2$/s/cm$^2$);
C, $O_2$ concentration at the gas interface (mon-$O_2$/cm$^3$);
$C_{cell}$, $O_2$ concentration at the cell layers (mon-$O_2$/cm$^3$);.
D, diffusion coefficient of oxygen (cm$^2$/s);
z, depth of diffusion path (cm); $\sigma$, surface cell density (cells/cm$^2$);
a, cellular oxygen consumption (mol-$O_2$/s/cell)

Fig. 2. Schematic illustration of oxygen diffusion and consumption in monolayers of adult rat hepatocytes cultured at confluence on conventional tissue culture-treated polystyrene (TCPS) and oxygen-permeable polydimethylsiloxane (PDMS) membranes in a steady state (**a**) and the simple equation describing the mass balance between oxygen supply and consumption (**b**). The estimated $O_2$ concentration at the cell layers can be calculated based on the equality of the oxygen diffusion flux driven by the concentration gradient and the oxygen consumption rate by the cells as written in equation (**b**).

(21, 22), but they applied the membrane to perfused bioreactors, thus complicating the culture format. In addition, there has been no article that understood the ultimate limitation of such membrane-based direct oxygenation in hepatocyte culture, particularly in terms of its feasibility in forming thick liver tissues.

We therefore used PDMS as another oxygen-permeable material, prepared a special microplate format, and investigated its feasibility in thick liver tissue engineering. When we use a PDMS membrane of 1-mm thickness and directly supply the cells with oxygen from the bottom surface to which they are attached, maximal oxygen supply (supposing that the oxygen concentration becomes 0 mol $O_2$/cm$^3$ at the cell layer) well exceeds the cellular demand (Fig. 2). In addition, the actual oxygen concentration at the cell layer is very close to the concentration that is decided by Henry's law due to the very low transport resistance of the membrane. This is likely to mimic the in vivo RBC-based oxygenation system, that is, completely meeting the tissue oxygen demands at a physiological low oxygen concentration, which may avoid excess oxidation stress to the tissues.

**5.2. Preparation of Special Microplates Equipped with PDMS Membranes**

PDMS membranes were prepared from a 10:1 mass ratio mixture of PDMS prepolymer and curing agent which was poured inside a large polystyrene box, degassed in a vacuum chamber, then cured for 2 h at 75°C in an oven. The PDMS membrane (1-mm thick) was stacked and clamped between the polycarbonate frame and stainless-steel board (Fig. 3) (17). The surface of the PDMS was treated with oxygen plasma using a reactive ion etching (RIE) apparatus. By this treatment, passive adsorption of type I collagen molecules in the solution onto PDMS surfaces was greatly enhanced and effective in stably maintaining Hep G2 cells on the surface. However, for adult and fetal hepatocytes, due to their high organization capability, covalent immobilization of collagen molecules was necessary. In that case, plasma-treated PDMS was further coupled with aminosilane and the induced amino groups were reacted with a photo-reactive cross-linker, SAND (Pierce, P-C21549; USA) (17, 23), by exposing them to UV light, and finally coated with a type I collagen solution. XPS analyses with and without detergent rinsing showed that collagen molecules were covalently immobilized onto the prepared PDMS surface.

**5.3. Complete Double-Layered Culture of Hepatocytes and Fibroblasts on PDMS Membranes**

Enhancement of oxygen supply enabled new types of coculture of rat hepatocytes with other cells of nonparenchymal origin such as NIH 3T3 fibroblasts (12). On collagen-coated TCPS surfaces, when we inoculate the two cells with a certain time interval, hepatocytes and fibroblasts never form complete double layers. Instead, hepatocytes form island-like structures surrounded by fibroblasts on the same TCPS surface. However, in the modified PDMS surfaces (with covalently immobilized collagen molecules), the two cell populations formed complete double layers according to the order of the inoculation, enabling heterogenic cell-to-cell direct interactions at the individual cell level (Fig. 4).

Surprisingly, this coculture remarkably enhanced the albumin production (almost 20 times) and its duration when compared to the conventional cocultures on TCPS (Fig. 5). Even after 2 weeks of culture, differences (though not so large) were observed in the albumin production, ammonium removal, and urea synthesis, which were enhanced between three- and fivefold compared to cocultures on TCPS (data not shown). In addition to hepatocyte functionality, regarding the normal polarity of hepatocytes, we recently reported that a combination of sandwich collagen culture and PDMS-based direct oxygenation remarkably enhanced the formation of bile canaliculi in hepatocyte monolayers (24). Such cellular behaviors resulting from the removal of the oxygen limitation can be expected to dramatically improve in vitro culture of hepatocytes in the near future.

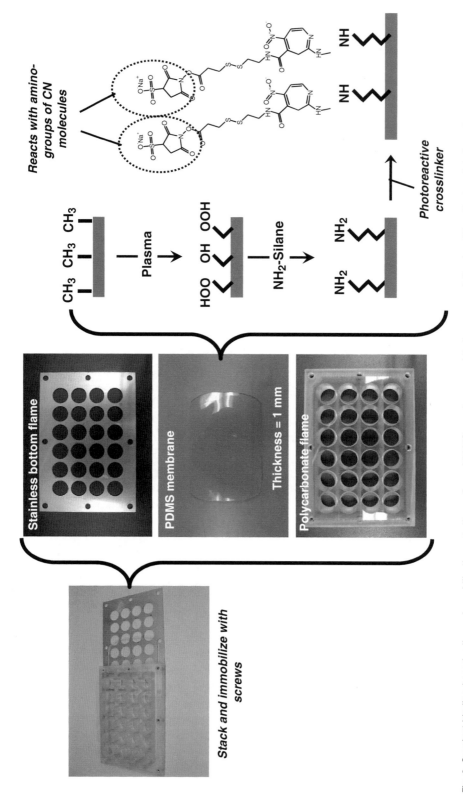

Fig. 3. Covalent binding-based collagen-immobilization onto the surface of PDMS membranes and their accommodation to a 24-well microplate format.

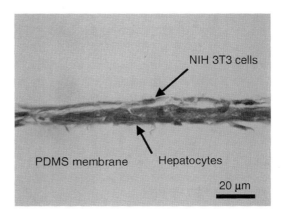

Fig. 4. Representative HE staining of a vertical thin section of a complete double-layered coculture on PDMS (day 13) (3T3 cells on top and hepatocytes below).

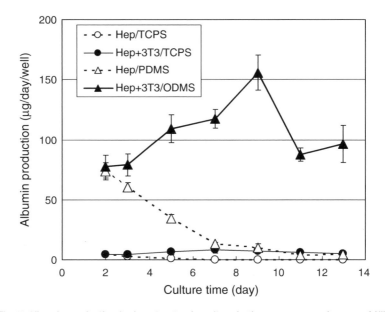

Fig. 5. Albumin production by hepatocytes in culture in the presence or absence of NIH 3T3 cells in TCPS (without oxygenation from below) and PDMS membrane-based plates (with direct oxygenation from below).

**5.4. Extensive Proliferation of Human Hepatoma Hep G2 Cells on PDMS Membranes**

To determine the limitation of such direct oxygenation culture using PDMS, we next employed proliferative hepatocarcinoma Hep G2 cells (25). As expected, the proliferation of Hep G2 cells was markedly enhanced, leading to the formation of a thick 3D cellular multilayer composed of 5–6 cell layers (Fig. 6). This was supported by the oxygen concentration profiles in the vicinity of

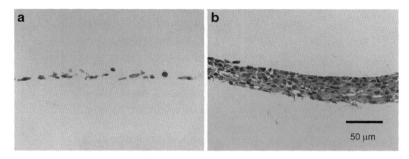

Fig. 6. HE staining of vertical thin sections of Hep G2 cells cultured on PDMS without oxygen supply from the bottom (**a**) and PDMS with direct oxygenation (**b**) on day 15.

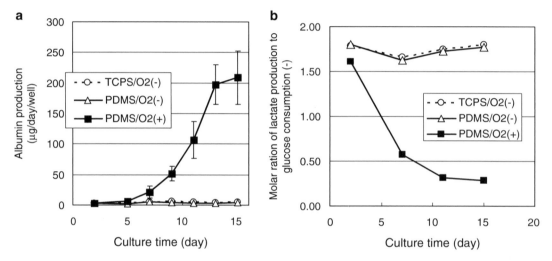

Fig. 7. Kinetics of albumin production (**a**) and molar ratio of lactate production/glucose consumption (**b**) during 15 days of culture. Mean ± SD (*N* = 6).

the cell layers, as predicted by numerical simulations. Note that plasma-treatment and collagen adsorption were enough to retain Hep G2 cells onto the PDMS surface. The cells also displayed a remarkably increased albumin production, reaching after 15 days 50 times that of cells cultured in conventional TCPS or the same treated PDMS surfaces without oxygen supply (Fig. 7a).

Very interestingly, the respiration profile in terms of glucose consumption and lactate production of hepatoma cells revealed the dominance of aerobic metabolism, and the higher albumin production agreed with the profile. When we take the molar ratios of produced lactate against consumed glucose, the stoichiometry ratios should be 2 for complete anaerobic respiration (glycolysis) and 0 for complete aerobic respiration (26). Under direct oxygenation using PDMS, the ratio decreased over time to about 0.30 but conventional submerged culture, where oxygen is supplied from the air–liquid interface, resulted in an over 1.60 ratio throughout the culture (Fig. 7b). This measurement clearly showed that PDMS-based direct oxygenation, although they formed thicker cell layers, enables the cells to utilize their aerobic energy production system that produces almost 20 times more ATP per amount of glucose (26). This agrees well with the very high albumin production of the cells (Fig. 7a). We also performed hypoxia inducible factor-1 (HIF-1) immunostaining and confirmed that oxidative stress was suppressed in the PDMS-based direct oxygenation culture in contrast to the conventional nonoxygenated cultures (data not shown). This suggests that the PDMS-based culture realizes an in vivo-like oxygenation situation, meeting the demand with low oxidative stress.

### 5.5. Spontaneous Formation of Heterogenic Liver Tissues from Fetal Rat Liver Cell Populations on a PDMS Membrane

We further checked the feasibility of the PDMS-based direct oxygenation system in fetal rat liver cells, because it is likely to be a very good model for hepatocyte progenitors, who are likely to be used in liver tissue engineering. Considering the low resistance of fetal cells to oxidative stress, we cultured the cells under different oxygen conditions: 5 or 21% oxygen for 2 or 1 week under 5% oxygen followed by 1 week under 21% oxygen. Due to the very high reorganization capability of fetal cells even in vitro, covalent binding of collagen molecules was indispensable to stably attach the cells to the PDMS (14).

As partly expected from the behaviors of proliferative Hep G2 cells, rat fetal liver cells were able to organize themselves into thick tissues, but their structures were more complex, composed of an epithelium of hepatocytes above mesenchyme-like tissues (Fig. 8). The thickness of this lower supportive mesenchymal tissue was directly correlated to atmospheric oxygen concentrations and was higher under 5% oxygen than under 21% (Fig. 8a, b). Interestingly, when cultures were switched after 1 week from 5 to 21% oxygen, we observed that lumen-containing structures were formed into thick mesenchymal-like tissue (Fig. 8c). Albumin and CK18 immunostaining revealed both the top epithelial cells in the three groups and the cells forming lumen-like structures in the switched oxygen culture (Fig. 8c) were all hepatocytes.

As expected, the improved oxygen supply using PDMS enhanced the albumin secretion rate, which was higher under higher oxygen concentration (Fig. 9a). Interestingly, the switched oxygen concentration culture group (5 → 21%) in PDMS showed a continuous increase in albumin production as opposed to the other three groups, whereas the 21% oxygen culture group showed initially increased albumin production that subsequently declined. This was in good agreement with the number of hepatocytes after 2 weeks of culture as evidenced from the vertical

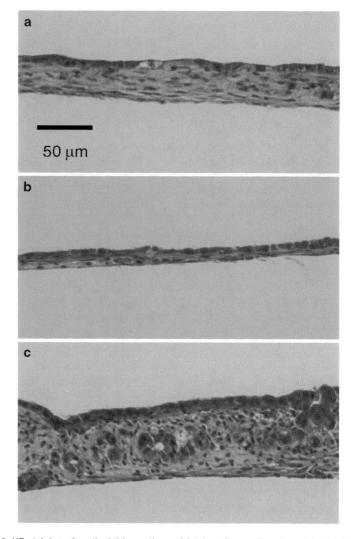

Fig. 8. HE staining of vertical thin sections of fetal rat liver cells cultured for 14 days on PDMS with direct oxygenation under 5% (**a**), 21% (**b**), and 5–21% (**c**) (switched on day 8) of atmospheric oxygen.

cross-sections (Fig. 8). It seems that better development of lower mesenchymal cell layers enables a sustained functional increase of fetal hepatocytes. Measurements of respiration profiles showed that the PDMS-based direct oxygenation culture enhanced the aerobic respiration (Fig. 9b). This should enable highly efficient energy production of the cells and seems to partly contribute to the extensive reorganization of hepatic tissues particularly under the switched oxygen condition.

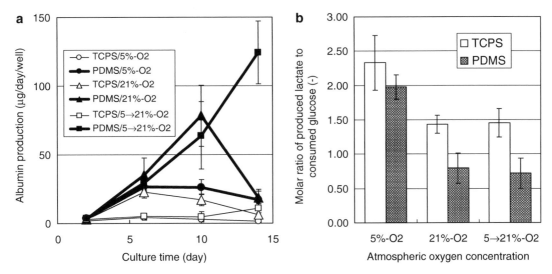

Fig. 9. Albumin production for 14 days of culture (**a**) and molar ratio of produced lactate to consumed glucose (**b**) on day 14. Mean ± SD (*N* = 6).

## 6. In Vitro and In Vivo Behaviors of Hepatocyte Progenitors Immobilized in Macroporous 3D Scaffolds

### 6.1. Superiority of Fetal Hepatocytes Over Adult Hepatocytes

3D macroporous sheet-based liver tissue engineering rather follows the conventional methodology of tissue engineering. However, as pointed out by Mooney's group in their excellent latest transplantation experiment using adult hepatocytes with local delivery of various growth factors, there seems to be a limitation in terms of long-term functionality when we use mature hepatocytes with a low proliferative capability (9). In contrast, we have been culturing fetal hepatocytes from mice, rats, and pigs in 3D macroporous scaffolds with appropriate growth factor cocktails and demonstrated that 3D culture with an appropriate growth factor cocktail enables remarkable functional maturation of fetal hepatocytes to at nearly mature functional levels. In addition, such fetal hepatocytes precultured in 3D scaffolds can further grow upon implantation to animals. This is a distinctive difference with the behaviors of adult hepatocytes as described above. We recently wrote a book chapter describing the results in detail and briefly summarize here the results both in vitro and in vivo (27).

### 6.2. Fabrication of PLLA-Based 3D Macroporous Scaffolds and Their Use in Culture (13)

The scaffolds were prepared using gas foaming salt as a porogen additive, as reported by Nam et al. (28). After dissolving PLLA at 80 mg/mL in chloroform, 1.5 g of $NH_4HCO_3$ particles per mL chloroform was added and mixed thoroughly. The highly viscous $PLLA/NH_4HCO_3$ in chloroform was put into Teflon molds with an inner diameter of 10 mm, and allowed to dry at 50–60°C for

15 min. We carefully removed it from the mold and dried it at 50–60°C for 30 min to let the chloroform evaporate completely. The PLLA/NH$_4$HCO$_3$ was cut into about 1.2-mm thick discs, so that the volume of each disc was about 0.10 cm$^3$. The discs were put into distilled water heated constantly to about 80°C and stirred for 30 min. They were then put into 70% ethanol for 5 min. We repeated this washout procedure at least three times until all the NH$_4$HCO$_3$ was completely removed from the discs. The prepared macroporous PLLA scaffolds were coated with type-I collagen and equilibrated with FBS-containing culture medium prior to cell inoculation. Cells were inoculated at an appropriate density to the disc put in a six-well plate (bacterial grade or suspension cell culture grade) with 2-mL culture medium and continuously shaken at 60–70 rpm on a rotational shaker in an incubator.

### 6.3. Efficiency In Vitro and In Vivo

In such 3D microenvironments, inoculated cells were organized into heterogenic 3D aggregates or multilayers, and their functions and in vitro stability were greatly enhanced when compared with those in 2D monolayer cultures. Although the detoxification capacity in terms of EROD measurement did not seem to be fully matured, other typical functions such as albumin production reached adult levels. This was enabled by the synergistic effects of 3D culture and soluble factor cocktails. Combination of nicotinamide (NA), dimethylsulfoxide (DMSO), and OSM was very effective in fetal mice culture (16), but does not support the growth and maturation of fetal rat hepatocytes. Based upon the cocktail effective for hepatic differentiation of embryonic stem cells (29), a cocktail composed of NA, HGF, FGF-1, FGF-4, OSM, and sodium butyrate, was very effective in the maturation of fetal rat liver cells (13). In the case of fetal porcine hepatocytes, presumably because the obtained hepatocytes were in a more advanced stage of maturation than mice and rats, dependency on soluble factors was low and 3D culture itself remarkably enhanced their spontaneous growth and maturation (15).

Although the combination of 3D culture and growth factor cocktails provides a suitable microenvironment for the growth and maturation of fetal hepatocytes, the biggest problem is the low final growth ratio (several times the inoculum, at most) and density (less than one tenth that that of in vivo liver tissues). This is because the cellular growth is mostly limited to a certain distance from the inner surface of the macropores, even with thin disk-shaped scaffolds and with continuous shaking. This indicated the insufficient mass transfer (primarily of oxygen) between culture medium and the inner spaces of the scaffolds.

However, this limitation can largely be overcome by in vivo implantation to the mesentery leaves of animals. Upon implantation, cells that had stopped their growth in vitro again started to proliferate, so that they finally filled the remaining macroporous spaces (Fig. 10) (27). This was partly attributed to the intrusion of microvessels from the surrounding tissues (30). These results clearly demonstrate that fetal cells precultured in appropriate 3D microenvironments can be expected to partly support the insufficient host liver functionality upon implantation.

## 7. Engineering of 3D Liver Tissues Using Special Scaffolds Comprising an Artificial Vasculature

### 7.1. Approaches for Engineering Vascularized Liver Tissues

To engineer larger 3D liver tissue, supply of blood or culture medium flow is definitely necessary. The general vascularization strategies in tissue engineering were lately summarized in several reviews including ours (5, 6, 10). The main in vitro strategies from an engineering point of view are (1) microelectromechanical systems (MEMS)-related approach – complete 2D fabrication of microfluidic vascular networks that supply adjacent hepatocytes; those 2D constructs are then stacked to produce 3D thick tissues (6, 31–33); this approach enables control of the flow within the construct since the flow is primarily determined by the geometry; however, the hepatocyte cell mass of current constructs is relatively low and scaling-up remains difficult; (2) modular assembly – tissue elements covered with endothelial cells are packed in some 3D space and the gaps finally generated between the elements is supposed to act as a macroscale flow channel network (34, 35); this approach is very promising but constructs, which are both perfusable and highly cell-dense, have not yet been produced; (3) 3D microfabrication-related approach – flow channels are fabricated in 3D macroporous scaffolds so that cells are stably immobilized in the macroporous part of the scaffolds and the culture medium preferentially flows in the channels due to their lower pressure drop (10).

Overall, there is no single approach that can achieve both tissue microstructure and clinically significant tissue mass at present. We have been employing the third (3D microfabrication-related) approach, because this approach is based on the most reliable fabrication technology at present. We proposed a special 3D design with flow channels, which branch and join in a smooth manner, thus forming a tree-like network within the macroporous structure, so that the network has one inlet and one outlet but the entire scaffold can be uniformly supplied with culture medium or blood perfusion (18) (Fig. 11a, b). To simply express such a kind of concept, we used tetrahedrons as a unit structure to compose the entire scaffold with a total volume of 13 cm³. In this design, oxygen is supplied to the cells in the macroporous structure from the flow channels and the maximum allowable edge length of the unit tetrahedron can be determined by one boundary condition: the oxygen concentration at the center of the space (composed of

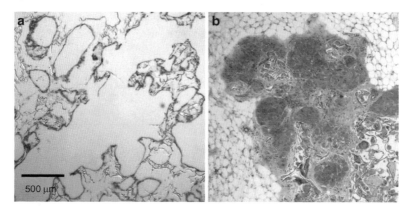

Fig. 10. HE staining of thin sections of fetal mouse hepatocyte-loaded macroporous PLLA scaffolds recovered after 1 week of in vitro culture (**a**) and after subsequent 4 weeks of implantation in mice (**b**).

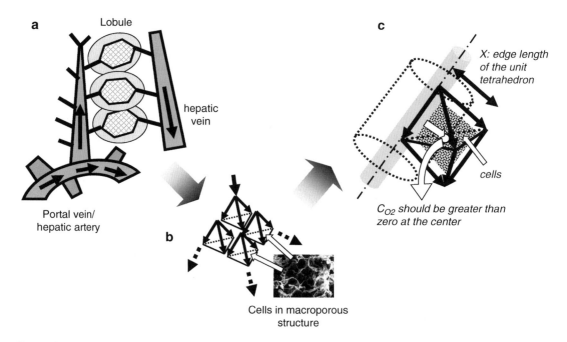

Fig. 11. Simple design of macroscale vasculature-like artificial flow channels based on oxygen diffusion and consumption around the channels by cells immobilized in the macroporous structure. (**a**) Vasculature in the liver; (**b**) simplified modeling of the liver vasculature using stacked unit tetrahedrons. The slanting edges are used as flow channels and the inner macroporous spaces are used for cell growth; (**c**) determination of $X$, the edge length of the unit tetrahedron according to the oxygen diffusion and consumption by the cells present around the flow channels.

two tetrahedrons, Fig. 11c) should always be higher than zero to ensure cellular viability. A simple equation describing the decrease in oxygen concentration around a single flow channel based on oxygen diffusion and consumption by the cells (36) predicted that the edge length of the unit tetrahedrons should be less than 200–400 μm if we use normal hepatocytes or hepatoma cells (Fig. 12).

**a**

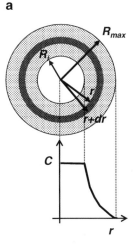

Basic differential equation;

$$\frac{d^2C}{dr^2} + \frac{1}{r}\frac{dC}{dr} - \frac{\rho\,a}{D} = 0$$

Boundary conditions;

$$C(r = R_i) = C_0$$

$$\left.\frac{dC}{dr}\right|_{r=R_{max}} = 0$$

$$C(r = R_{max}) = 0$$

C, oxygen concentration; r, radius; $\rho$, cell density, a, cellular oxygen consumption rate; D, diffusion coefficient in cell-loaded macroporous scaffolds.

Analytical solution;   $C = C_0 + \dfrac{\rho\,a}{2D}\left[\left(\dfrac{r^2}{2} - \dfrac{R_i^2}{2}\right) + R_{max}^2\left(\ln\dfrac{R_i}{r}\right)\right]$

**b**

**Flow channel, φ200 μm**

X: Edge of the tetrahedron (μm)

Cell density (cells/cm³)

In vivo cell density in the liver tissue density

Fig. 12. Theoretical calculation of the oxygen concentration profile produced around the flow channels (a) and relation between the maximum allowable X value and the intended cell density in the macroporous scaffolds (b).

### 7.2. Fabrication of 3D Scaffolds Using the Selective Laser Sintering Process

Unfortunately, we faced a technological limitation of the current 3D fabrication process: the difficulty in obtaining completely penetrating and slanting channels when fabricating a macroporous structure using salt particles as a porogen (37). The resolution of the 3D fabrication process is still around several hundred micrometers, almost 1,000 times lower than MEMS-based 2D micropatterning methods, and it is inherently difficult to fabricate such slanting channels in such layer-by-layer 3D fabrication. Therefore, we had no choice but to enlarge the size of the unit tetrahedron to 4 mm in its edge length, which was mainly decided by the fact that the allowable diameter of the slanting flow channel in our selective laser sintering (SLS) machine was 0.8 mm. Anyway, a PCL scaffold with an overall porosity of 89% was obtained after water leaching of porogen salt particles with the help of water perfusion of the flow channel network (Fig. 13a–d) (18).

### 7.3. Perfusion Culture of Hepatocarcinoma Hep G2 Cells and Fetal Porcine Hepatocytes

For perfusion culture, we connected two silicone tubes to the inlet and outlet and completely covered the outer surface of the scaffolds with silicone resin. Culture of Hep G2 cells clearly showed enhanced growth and functions by arranging such a network; no substantial growth was obtained in a channel-free scaffold (Fig. 14). In addition, initial stable cell attachment using avidin–biotin binding (38) strongly influenced the final growth. As expected, cells mainly distributed within 200 μm from the channel and the overall cell density was $2 \times 10^7$ cells/cm³ scaffold (Fig. 13e). This experimental observation clearly shows that if we can fabricate an improved

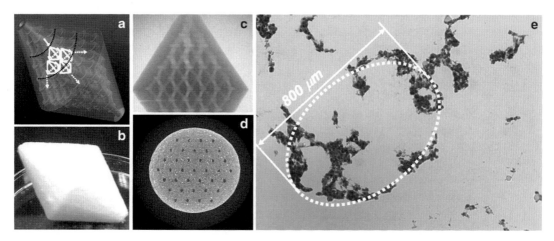

Fig. 13. Fabricated 3D scaffolds and HE thin section on day 9 of a Hep G2 perfusion culture. (**a**) CAD design of 13 cm³ PCL macroporous scaffold composed of tetrahedrons whose edge length is 4 mm and flow channels 2, 1.5, and 1 mm in diameter according to their position in the scaffold; (**b**) appearance of the PCL scaffold; (**c, d**) micro-X ray CT pictures of vertical and horizontal section of the scaffold; (**e**) HE thin section of a cell-loaded scaffold, possibly showing the cell growth around flow channels.

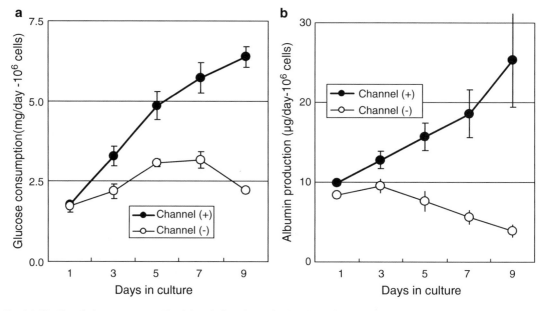

Fig. 14. Kinetics of glucose consumption (**a**) and albumin production (**b**) during 9 days of perfusion culture of Hep G2 cells in scaffolds with or without flow channels. Consumption and production are shown per 10⁶ cells immobilized on day 1.

scaffold in which all the cells can be immobilized within 200 μm from the channel, the overall density can reach the same as that of in vivo liver ($2.5 \times 10^8$ cells/cm³ scaffold). We also obtained good cellular growth and increase in albumin production of fetal porcine hepatocytes (Fig. 15), but the final cell density was again a lower value, $0.85 \times 10^7$ cells/cm³ scaffold. These results strongly show

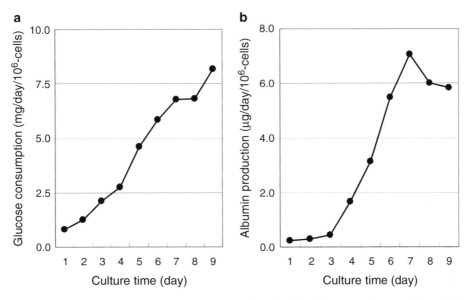

Fig. 15. Kinetics of glucose consumption (**a**) and albumin production (**b**) during 9 days of perfusion culture of fetal porcine hepatocytes in scaffolds with flow channels. Consumption and production are shown per $10^6$ cells immobilized on day 1.

the necessity to improve the resolution of fabrication processes. A recent advance using a special microsyringe disposition system enabled enhanced resolutions down to several tens of micrometers (39). Such improvements are likely to enable the direct and organized fabrication of the macroporous structure itself, as opposed to the current random process using porogens, in the near future.

## 8. Conclusions

In this chapter, we introduced ours and related approaches to liver tissue engineering in three configurations such as thick cellular sheets, macroporous scaffold sheets, and flow channel-containing macroporous scaffolds (Fig. 1). Under the current boundary condition that in vitro prevascularization of the engineered tissue is almost impossible, the former two configurations are promising in actual human clinical trials for liver diseases that can be treated by implantation of relatively small tissue constructs. The third approach should enable the implantation of liver tissues of a much larger mass in the future. In any case, we would like to stress the fact that oxygen supply is the key factor to design and organize liver tissue. This enables the cells to utilize aerobic respiration that produces almost 20 times more ATP than anaerobic respiration for the same glucose consumption. This also allows the cells to use their maximum reorganization capability that cannot be observed in conventional anaerobic conditions. One thing we need to consider is the actual oxygen concentration at the cell surface to avoid excess oxidative stress. This seems to be very important when we try to obtain mature hepatocytes from stem or progenitor cells. The following are remaining important issues to be overcome in the future.

## 9. Expected Future Works

### 9.1. Formation of a Bile Canaliculi Network

In current engineered liver tissues of small mass, bile acids produced by the cells and accumulated in the bile canaliculi leak back to the blood flow before being finally eliminated by the remaining host liver through the host bile duct (3). However, if we really think about substitution of the host liver with engineered liver tissues, we need to arrange a bile canaliculi/duct network over the engineered tissues. One clue to the solution is the experimental results by Sudo et al. (40, 41) about the functional canaliculi network and transport of bile acids to small bile pools formed in hepatocyte progenitor colonies. As a next trial, combination with appropriate microfabrication technologies may give a new insight into organization of such advanced hepatic tissues in vitro.

### 9.2. Endothelialization and Angiogenesis

For the third approach where a macroscale flow channel network is to be arranged, because it is finally perfused with the host blood flow upon implantation, complete pre-endothelialization of all the inner surfaces are necessary. In addition, further angiogenesis toward the macroporous structure is expected to allow good mass transfer between the cells and the blood flow. At present, no one has succeeded in vitro in either complete endothelialization of engineered tissues or formation of a perfusable microvasculature. Overall, recent reports in this area demonstrate the necessity of various supporting cells such as fibroblasts or pericytes (or their progenitors) as well as the parenchymal cells of the relevant organ, since those supporting cells promote not only vascular formation (42) but also liver progenitor maturation (43).

In vivo, the first system that forms in the embryo is the vascular system, and other organs develop subsequently; it has been shown that liver and pancreas formation in the embryo is promoted by endothelial cells (44, 45). In future, the most efficient way to produce an artificial liver may be to reproduce in vitro the conditions that enable liver formation in the embryo, using embryonic stem cells or induced pluripotent stem cells in prevascularized scaffolds. In any case, developmental biology may provide important guiding principles to tissue engineers.

**9.3. Macroscale Oxygenation**

When we really target organization of large liver tissues, we need to pay attention to macroscale clean from oxygen depletion in addition to microscale considerations. This is the simple mass balance between tissue oxygen consumption and the oxygen amount supplied by the vessel. Very low oxygen solubility in culture medium is the first limiting factor; $2 \times 10^{-7}$ mol $O_2$/mL culture medium under 21% $O_2$ in the gas phase, which is about 1/70th that of blood. The second limiting factor is the maximum tolerable shear stress at the inner walls of the flow channels, ~15 dyn/cm$^2$, which restricts the maximum culture medium flow rate. When we think about a typical hepatocyte arrangement around a microvessel, the maximum number of viable hepatocytes along the vessel is calculated to be 24.5 cells (Fig. 16), leading to the prediction that the maximum feasible tissue volume is only 50 cm$^3$. Therefore, effective oxygen carriers should be incorporated into the culture medium when we intend to produce large liver tissue equivalents.

There are two types of oxygen carriers: perfluorocarbon (PFC)-based and hemoglobin (Hb)-based carriers. PFC is used as a suspension of its emulsions and the maximum concentration of PFC is around 15% (v/v). Therefore, the overall oxygen solubility of such PFC-containing culture medium is at most four times higher than that for culture medium. Among the various Hb-based RBC substitutes, polyethyleneglycol (PEG)-decorated liposome encapsulated Hb (LEH) is one of the most promising designs for a RBC

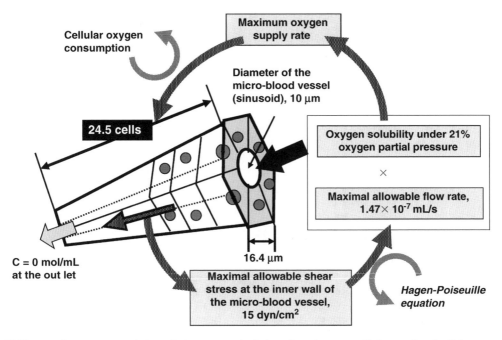

Fig. 16. Macroscale oxygen mass balance between per minute-based maximum supplied amount and cellular consumption along one blood vessel.

substitute for infusion to humans (46). We therefore checked its toxicity and efficacy in adult and fetal rat hepatocytes. Although there was no toxicity in adult rat hepatocytes (47) and the LEH remarkably improved hepatocyte viability and functions in perfusion culture, LEH showed strong toxicity toward fetal hepatocytes. LEH incorporated into cells was broken up and released free Hb molecules in the cells, probably causing toxicity via production of reactive oxygen species (ROS), against which fetal cells have not yet developed sufficient defense mechanisms (48). Therefore, Hb-based carriers with improved design will be necessary in the future.

## Acknowledgments

The studies were carried out based on various scientific grants, such as Grant-in-Aids for Scientific Research from the Ministry of Health, Labor and Welfare, and those from the Ministry of Education, Culture, Sports, Science and Technology, Japan. Some of the studies were performed in the framework of LIMMS (Laboratory for Integrated Micro-Mechatronic Systems), a joint laboratory between the CNRS (Centre National de la Recherche Scientifique) and IIS (Institute of Industrial Science), the University of Tokyo. We thank Saipaso Research Center (Tokyo, Japan) for preparing histological samples. F. Evenou and K. P. Montagne were supported by the Research Fellowship for Young Scientists from the Japan Society for the Promotion of Science.

## References

1. Demetrious AA, Jr. Brown RS, Busuttil RW, Fair J, McGuire BM, Rosenthal P, II Am Esch JS, Lerut J, Nyberg SL, Salizzoni M, Fagan EA, de Hamptinne B, Broelsch CE, Muraca M, Salmeron JM, Rablin JM, Metselaar HJ, Pratt D, DE La Mata M, McChesney LP, Everson GT, Lavin PT, Stevens AC, Pitkin Z, Solomon BA (2004) Prospective, randomized, multicenter, controlled trial of a bioartificial liver in treating acute liver failure, Ann Surg **239**, 660–667.

2. Fox IJ, Chowdhury JR, Kaufman SS, Goertzen TC, Chowdhury NR, Warkentin PI, Dorko K, Sauter BV, Strom SC (1998) Treatment of the Crigler-Najjar syndrome type I with hepatocyte transplantation, New Eng J Med., **338**, 1422–1426.

3. Ohashi K (2008) Liver tissue engineering: The future of liver therapeutics, Hepatol Res, **38**, S1, S76-87.

4. Fiegel HC, Kaufmann PM, Bruns H, Kluth D, Horch RE, Vacanti JP, Kneser U (2008) Hepatic tissue engineering: from transplantation to customized cell-based liver directed therapies from the laboratory. J Cell Mol Med, **12**, 56–66.

5. Lovett M, Lee K, Edwards A, Kaplan DL (2009) Vascularization strategies for tissue engineering, Tissue Eng B, **15**, 353–370.

6. Hoganson DM, Pryor II HI., Vacanti JP (2008) Tissue engineering and organ structure: a vascularized approach to liver and lung, Pediat Res, **63**, 520–526.

7. Ohashi K, Yokoyama T, Yamato M, Kuge H, Kanehiro H, Tsutsumi M, Amanuma T, Iwata H, Yang J, Okano T, Nakajima Y (2007) Engineering functional two- and three-dimensional liver systems in vivo using hepatic tissue sheets, Nat Med, **13**, 880–885.

8. Mooney DJ, Kaufmann PM, Sano K, McNamara KM, Vacanti JP, Langer R (1994), Transplantation of hepatocytes using porous biodegradable sponge, Transplant Proc, **26**, 3425–3426.

9. Smith MK, Riddle KW, Mooney DJ (2006), Delivery of heterotrophic factors fails to enhance longer-term survival of subcutaneously transplanted hepatocytes, Tissue Eng, **12**, 235–244.

10. Sakai Y, Huang H, Hanada S, Niino T (2010) Toward engineering of vascularized three-dimensional liver tissue equivalents possessing a clinically-significant mass, Biochem Eng J, **48**, 348–361.

11. Seglen PO (1976) Preparation of isolated liver cells. In Methods in Cell Biology, vol. 13 (Etd. by Prescott, D. M.), pp. 29–83, Academic Press, New York.

12. Nishikawa M, Kojima N, Komori K, Yamamoto T, Fujii T, Sakai Y (2008) Enhanced maintenance and functions of rat hepatocytes induced by combination of on-site oxygenation and coculture with fibroblasts, J Biotechnol, 133, 253–260.

13. Hanada S, Kojima N, Sakai Y (2008) Soluble factor-dependent in vitro growth and maturation of rat fetal liver cells in a three-dimensional culture system, Tissue Eng A, **14**, 149–160.

14. Hamon M, Hanada S, Fujii T, Sakai Y (2011) Direct oxygen supply with polydimethylsiloxane (PDMS) membranes induces a spontaneous organization of thick heterogeneous liver tissues from rat fetal liver cells in vitro, Cell Transplant, in press.

15. Huang H, Hanada S, Kojima N, Sakai Y. (2006) Enhanced functional maturation of fetal porcine hepatocytes in three-dimensional poly-L-lactic acid scaffolds: a culture condition suitable for engineered liver tissues in large-scale animal studies, Cell Transplant, **15**, 799–809.

16. Jiang J, Kojima N, Guo L, Naruse K, Makuuchi M, Miyajima A, Yan W, Sakai Y (2004) Efficacy of engineered liver tissue based on poly-L-lactic acid scaffolds and fetal mouse liver cells cultured with oncostatin M, nicotinamide and dimethyl sulfoxide, Tissue Eng, **10**, 1577–1586.

17. Nishikawa M, Yamamoto T, Kojima N, Komori K, Fujii T, Sakai Y (2008) Stable immobilization of rat hepatocytes as hemispheroids onto collagen-conjugated poly-dimethylsiloxane (PDMS) surfaces: importance of direct oxygenation through PDMS for both formation and function, Biotechnol Bioeng, **99**, 1472–1481.

18. Huang H, Oizumi S, Kojima N, Niino T, Sakai Y (2007) Avidin-biotin binding–based cell seeding and perfusion culture of liver-derived cells in a porous scaffold with a three-dimensional interconnected flow-channel network, Biomat., **28**, 3815–3823.

19. Stevens KM (1965) Oxygen requirement for liver cells in vitro, Nature, **206**, 199.

20. Smith MD, Smirthwaite AD, Cairns DE, Cousins RB, Gaylor JD (1996) Techniques for measurement of oxygen consumption rates of hepatocytes during attachment and post-attachment. Int J Artif Organs, **19**, 36–44.

21. Tilles AW, Baskaran H, Roy P, Yarmush ML, Toner M (2001) Effects of oxygenation and flow on the viability and function of rat hepatocytes cocultured in a microchannel flat-plate bioreactor. Biotechnol Bioeng, **73**, 379–389.

22. de Bartolo L, Salerno S, Morelli S, Giorno L, Rende M, Memoli B, Procino A, Andreucci VE, Bader A, Drioli E (2006) Long-term maintenance of human hepatocytes in oxygen-permeable membrane bioreactor. Biomat, **27**, 4794–4803.

23. Wang N, Ostuni E, Whitesides GM, Ingber DE (2002) Micropatterning tractional forces in living cells. Cell Motil Cytoskeleton, **52**, 97–106.

24. Matsui H, Osada T, Moroshita Y, Sekijima M, Fujii T, Takeuchi S, Sakai Y (2010) Rapid and enhanced repolarization in sandwich-cultured hepatocytes on an oxygen-permeable membrane, Biochem Eng. J, **52**, 255–262.

25. Evenou F, Fujii T, Sakai Y (2010), Spontaneous formation of highly functional three-dimensional multilayer from human hepatoma Hep G2 cells cultured on an oxygen-permeable polydimethylsiloxane membrane, Tissue Eng C, **16**, 311–318.

26. Alberti KG (1977) The biochemical consequences of hypoxia, J Clin Pathol, Suppl (R Coll Pathol) **11**, 14–20.

27. Sakai Y, Jiang J, Hanada S, Huang H, Katsuda T, Kojima N, Teratani T, Ochiya T (2011) Three-dimensional culture of fetal mouse, rat and porcine hepatocytes, in "Fetal Tissue Transplantation", etd. by N. Bhattacharya and P. Stubblefield, Springer-Verlag UK, in press.

28. Nam YS, Yoon JJ, Park TG (2000) A novel fabrication method of macroporous biodegradable polymer scaffolds using gas foaming salt as a porogen additive. J Biomed Mater Res, **53**, 1–7.

29. Teratani T, Yamamoto H, Aoyagi K, Sasaki H, Asari A, Quinn G, Sasaki H, Terada M, Ochiya T (2005) Direct hepatic fate specification from mouse embryonic stem cells, Hepatol, **41**, 836–846.

30. Katsuda T, Teratani T, Ochiya T, Sakai Y (2010) Transplantation of a fetal liver cell-loaded hyaluronic acid sponge onto the mesentery

recovers a Wilson's disease model rat, J Biochem, **148**, 281–288.

31. Park J, Li Y, Berthiaume F, Toner M, Yarmush ML, Tilles AW (2008) Radial flow hepatocyte bioreactor using stacked microfabricated grooved substrates, Biotechnol Bioeng, **99**, 455–467.

32. Carraro A, Hsu WM, Kulig KM, Cheung WS, Miller ML, Weinberg EJ, Swart EF, Kaazempur-Mofrad M, Borenstein JT, Vacanti JP, Neville C (2008) In vitro analysis of a hepatic device with intrinsic microvascular-based channels, Biomed Microdevices, **10**, 795–805.

33. Hoganson DM, Pryor II HI, Spool ID, Burns OH, Gilmore JR, Vacanti JP (2010) Principles of biomimetic vascular network design applied to a tissue-engineered liver scaffold, Tissue Eng A, **16**, 1469–1477.

34. McGuigan AP, Sefton MY (2006) Vascularized organoid engineering by modular assembly enables blood perfusion., Proc Nat Acad Sci, **103**, 11461–11466.

35. Inamori M, Mizumoto H, Kajiwara T (2009) An approach for formation of vascularized liver tissue by endothelial cell–covered hepatocyte spheroid integration, Tissue Eng A, **15**, 2029–2037.

36. Krogh A (1919) The number and the distribution of capillaries in muscle with the calculation of the oxygen pressure necessary for supplying the tissue, J Physiol. (Lond) **52**, 409–515.

37. Niino T, Sakai Y, Huang H, Naruke H (2006) SLS fabrication of highly porous model including fine flow channel network aiming at regeneration of highly metabolic organs, Solid Freeform Fab. Proc., 160–170.

38. Kojima N, Matsuo T, Sakai Y (2006) Rapid hepatic cell attachment onto biodegradable polymer surfaces without toxicity using an avidin-biotin binding system, Biomat, **27**, 4904–4910.

39. Vozzi G, Previti A, De Rossi D, Ahluwalia A (2002) Microsyringe-based deposition of tow-dimensional and three-dimensional polymer scaffolds with a well-defined geometry for application to tissue engineering, Tissue Eng, **8**, 1089–1098.

40. Sudo R, Mitaka T, Ikeda S, Sugimoto S, Harada K, Hirata K, Tanishita K, Mochizuki Y (2004) Bile canalicular formation in hepatic organoid reconstructed by rat small hepatocytes and hepatic nonparenchymal cells, J Cell Physiol, **199** 252–261.

41. Sudo R, Kohara H, Mitaka T, Ikeda M, Tanishita K (2005) Coordination of bile canalicular contraction in hepatic organoid reconstructed by rat small hepatocytes and nonparenchymal cells. Ann Biomed Eng, **33**, 696–708.

42. Lokmic Z, Mitchell GM (2008) Engineering the microcirculation, Tissue Eng B, **14**, 87–103.

43. Wang Y, Yao HL, Cui CB, Wauthier E, Barbier C, Costello MJ, Moss N, Yamauchi M, Sricholpech M, Gerber D, Loboa EG, Reid LM (2010) Paracrine signals from mesenchymal cell populations govern the expansion and differentiation of human hepatic stem cells to adult liver fates, Hepatol, **52**, 1443–1454.

44. Matsumoto K, Yoshitomi H, Rossant J, Zaret KS (2001) Liver organogenesis promoted by endothelial cells prior to vascular function, Science, **294**, 559–563.

45. Lammert E, Cleaver O, Melton D (2001) Induction of pancreatic differentiation by signals from blood vessels, Science, **294**, 564–567.

46. Takahashi A (1995) Characterization of Neo Red Cell (NRCs), their function and safety; in vitro test, Artif. Cells Blood Subst Immob Biotechnol, **23**, 347–354.

47. Naruto H, Huang H, Nishikawa M, Kojima N, Mizuno A, Ohta K, Sakai Y (2007) Feasibility of direct oxygenation of primary-cultured rat hepatocytes using polyethylene glycol decorated liposome-encapsulated hemoglobin (LEH), J Biosci Bioeng, **104**, 343–346.

48. Montagne K, Huang H, Ohara K, Matsumoto K, Mizuno A, Ohta K, Sakai Y (2011) Use of liposome encapsulated hemoglobin as an oxygen carrier for fetal and adult rat liver cell culture, J Biosci Bioeng, in press.

# Chapter 17

# Mesenchymal Stem Cell Therapy on Murine Model of Nonalcoholic Steatohepatitis

## Yoshio Sakai and Shuichi Kaneko

## Abstract

A severely malfunctioning liver, due to acute liver injury or chronic liver disease, can lead to hepatic failure. The ultimate treatment for hepatic failure is liver transplantation; however, the availability of donors is a critical issue. Therefore, regenerative therapy is an anticipated novel approach for restoring liver function. Mesenchymal stem cells are pluripotent somatic cells that can differentiate into several cell types, including hepatocytes. Moreover, they are obtainable from easily accessible autologous adipose tissue, making them ideal for regenerative therapy. This chapter describes experimental methods for isolating mesenchymal stem cells from murine adipose tissues and expanding them, and also describes murine chronic liver disease, steatohepatitis, for the study of experimental regenerative treatments of chronic liver disease.

**Key words:** Mesenchymal stem cells, Adipose tissue, Nonalcoholic steatohepatitis

## 1. Introduction

Liver disease is a major health issue worldwide, and includes chronic hepatitis and acute liver failure due mostly to infection by hepatitis or other viruses and drug hepatotoxicity (1). The most intense form of acute liver injury is fulminant hepatitis, which results in rapid and massive destruction of hepatocytes, leading to acute hepatic failure. By contrast, the pathological features of chronic liver diseases are characterized by persistent hepatic inflammation and subsequent fibrotic change that distorts the fine lobular architecture of the liver tissue. This ultimately leads to end-stage chronic liver injury, which manifests clinically as encephalopathy, due to the failure of various metabolic processes and impaired portal circulation. The liver is unique in that hepatocytes per se (2), or progenitor cells (3, 4) can proliferate and restore the original architecture

Takahiro Ochiya (ed.), *Liver Stem Cells: Methods and Protocols*, Methods in Molecular Biology, vol. 826,
DOI 10.1007/978-1-61779-468-1_17, © Springer Science+Business Media, LLC 2012

and function of the liver. However, with massive destruction of parenchymal hepatocytes or chronic distortion of the liver architecture with advanced fibrosis the liver cannot regenerate sufficiently. The most effective and radical treatment for hepatic failure is liver transplantation. However, this is limited by the availability of donors, as there are too few donors compared to the population of hepatic failure patients. Even when a donor is available, the relatively high mortality of the transplantation procedure and the permanent requirement for immunosuppressants are major burdens to the recipient.

Regenerative therapy is a novel alternative treatment to liver transplantation for the severely impaired, malfunctioning cirrhotic liver. Bone-marrow stem cells are thought to contribute to liver regeneration (5–8), although it is controversial whether bone-marrow hematopoietic stem cells can differentiate into hepatocytes (9–12). Mesenchymal stem cells are pluripotent somatic stem cells that can differentiate into mesodermal lineage cells, such as adipocytes, chondrocytes, and osteocytes (13), as well as into nonmesodermal lineage cells, such as cardiomyocytes (14, 15) and hepatocytes (16–19). They reside in the bone marrow, umbilical cord, and adipose tissues; adipose tissues are especially rich in mesenchymal stem cells. For regenerative cell therapy, autologous cells would be ideal, avoiding the requirement for matching the major histocompatibility antigens to prevent immunological rejection. Consequently, bone marrow and adipose tissues are attractive sources of mesenchymal stem cells for regenerative therapy. Mesenchymal stem cells may also have favorable biological effects on fibrosis (20, 21) and inflammation (22). This chapter describes experimental methods for studying regenerative therapy for chronic liver disease using mesenchymal stem cells, the culture of mesenchymal stem cells from murine adipose tissue, and a murine model of steatohepatitis that resembles human nonalcoholic steatohepatitis (23). Other methods and their application to liver disease models are also discussed.

## 2. Materials

### 2.1. Reagents

1. Collagenase type I (Wako Pure Chemical Industries, Osaka, Japan).

2. Phosphate-buffered saline without calcium and magnesium [PBS(−)] (Wako Pure Chemical Industries).

3. DMEM/nutrient mixture Ham F-12 (DMEM/F12) with l-glutamine, 15 mM HEPES (Invitrogen, Life Technologies, Carlsbad, CA).

4. Fetal bovine serum (FBS, Invitrogen).

5. Antibiotic/antimycotic (100×), liquid (Invitrogen).

6. 0.05% w/v trypsin – 0.53 mmol/L EDTA-4Na (Wako Pure Chemical Industries).

7. Pentobarbital sodium (64.8 mg/ml) (Schering-Plough Animal Health, Tokyo, Japan).

8. Atherogenic and high-fat diet (ATH + HF): 38.25% CRF-1 (standard chow, Charles River Laboratories Japan, Yokohama, Japan), 60.0% cocoa butter, 1.25% cholesterol, 0.50% cholate (Oriental Yeast, Tokyo, Japan).

9. Ethanol.

10. α-Cyanoacrylate adhesive.

### 2.2. Reagent Preparation

1. Collagenase solution: 1 g of collagenase type I powder is dissolved in 133 ml of PBS and stored at −80°C until use.

2. Culture medium: DMEM/F12 supplemented with antibiotic/antimycotic liquid and 10% heat-inactivated FBS and stored at 4°C.

3. Dilute pentobarbital with PBS(−) at tenfold for anesthesia of mice.

### 2.3. Animal

C57Bl/6 J mice (male, 8–10 weeks old, Charles River Laboratories, Yokohama, Japan).

### 2.4. Equipment

1. Operating scissors.

2. Tweezers.

3. Needle.

4. Needle holder.

5. 15-ml polypropylene conical tube (BD Falcon, Franklin Lakes, NJ).

6. 100-μm cell strainer (BD Falcon).

7. 6-cm culture dish (Nunc, Rockside, Denmark).

8. 5-0 silk thread (Niccho Industry, Tokyo, Japan).

9. 27-gauge needle with 1-ml syringe.

## 3. Methods

### 3.1. Isolation and Culture of Murine Mesenchymal Stem Cells from Adipose Tissue

1. All animal experiments should comply with national laws and institutional regulations.

2. Euthanize a C57Bl/6J mouse by cervical dislocation.

3. Disinfect the skin with 70% ethanol.

4. Make a midline abdominal skin incision and peel off the skin to expose the subcutaneous inguinal region.

5. Obtain adipose tissue from the subcutaneous inguinal region by cutting the connective tissues between the adipose tissue and skin, and place in a 6-cm culture dish with PBS(–).

6. Remove the lymph nodes from the adipose tissue using tweezers (see Note 1).

7. Cut the obtained adipose tissue into 1–2-mm pieces with scissors.

8. Put the fragmented adipose tissue in a 15-ml conical tube containing 10 ml of PBS.

9. Centrifuge it at $200 \times g$ for 3 min and remove the supernatant.

10. Add 10 ml of PBS (–) to the tube and centrifuge it at $200 \times g$ for 3 min.

11. Remove the supernatant as in step 8.

12. Put the PBS(–)-rinsed adipose tissue fragments into a 15-ml conical tube with 2–3 ml of collagenase aliquot.

13. Incubate the adipose tissue fragments and collagenase with shaking at 37°C in thermostat bath for 1 h.

14. Add an equal volume of DMEM/F12 containing 10% heat-inactivated FBS supplemented with 1% antibiotic/antimycotic liquid.

15. Centrifuge at $200 \times g$ for 10 min.

16. Remove the debris and PBS(–).

17. Resuspend the remaining cells in PBS(–) and filter them through a 100-μm cell strainer.

18. Centrifuge at $200 \times g$ for 10 min.

19. Remove the PBS(–), suspend the cells in 4 ml of DMEM-F12 supplemented with heat-inactivated FBS, and place in a 6-cm culture dish (Fig. 1a).

20. Replenish the culture medium with fresh complete medium the next day.

21. Replenish the medium every 3–4 days. The culture usually reaches 70% cell confluence after 10 days (Fig. 1a).

22. Cells can usually be passaged and expanded eight or nine times until morphological change appears (Fig. 1b) (see Notes 2 and 3)

### 3.2. Establishing a Murine Steatohepatitis Model

C57Bl/6J male mice are maintained in colony cages with a 12-h light/12-h dark cycle. 8-week-old mice are fed an ATH+HF diet for 24 weeks. The livers of these mice develop steatosis in hepatocytes accompanied with pericellular fibrosis (Fig. 2a, b), resembling the liver histology seen in advanced nonalcoholic steatohepatitis (23).

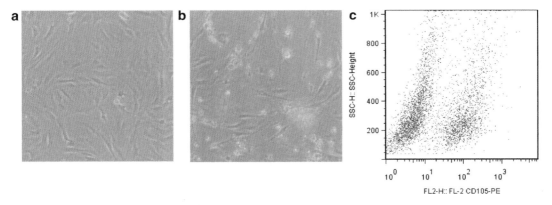

Fig. 1. The appearance of cultured cells obtained and expanded from murine adipose tissues. (**a**) The characteristic "spindle shape" of the mesenchymal stem cells is observed. (**b**) Morphological change of cells appeared usually after ten passages. (**c**) CD105 expression of cultured cells (eight passages).

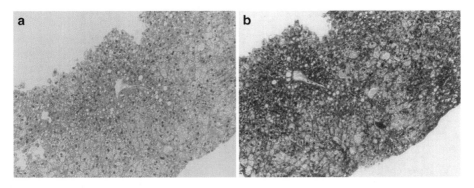

Fig. 2. Histology of the liver obtained from mice, which were fed with ATH + HF diet for 24 weeks. (**a**) HE staining (×100), (**b**) AZAN staining (×100).

**3.3. Experimental Therapeutic Application of Mesenchymal Stem Cells in the Murine Steatohepatitis Model**

1. Mice that develop steatohepatitis on the ATH + HF diet for 24 weeks are anesthetized by an intraperitoneal injection of 200 µl of diluted pentobarbital.

2. Expanded mesenchymal stem cells isolated from murine adipose tissues are prepared.

3. A midabdominal incision is made and the middle lobe of the liver is exposed.

4. A 2–3-mm liver specimen is obtained by cutting with scissors and the cut area is closed using α-cyanoacrylate adhesive.

5. After the biopsy, a $1 \times 10^5/200$ µl mesenchymal stem cell aliquot is injected into the subcapsule of the spleen using a 27-gauge needle with a 1-ml syringe.

6. The peritoneum and skin are sutured with 5-0 silk.

7. The mice are kept on the ATH + HF diet for two more weeks.

8. After 2 weeks, the mice are euthanized by cervical dislocation. Serum and liver tissues are collected, and RNA is extracted from the liver tissues. These samples are assayed to assess the therapeutic effect of administering mesenchymal stem cells (see Notes 4 and 5).

## 4. Notes

1. This step is required to avoid contamination of mesenchymal stem cell culture by resident lymphocytes.

2. The method of isolating and culturing mesenchymal stem cells from adipose tissues is described. Mesenchymal stem cells also reside in bone marrow, and the methods for isolating and expanding mesenchymal stem cells from bone marrow tissues have been reported (24, 25). The latter report states that mouse mesenchymal stem cells were isolated by aspiration of bone marrow in the tibia and femur, and cultured in DMEM supplemented with 15% FBS. The culture medium was replenished frequently. With this method, confluent mesenchymal stem cells can be obtained after 21 days.

3. CD105 is a marker for mesenchymal stem cells. Using the CD105 MultiSort Kit (Miltenyi, Auburn, CA), the mesenchymal cell fraction can be enriched.

4. To assess the effect of mesenchymal stem cells on the liver in the steatohepatitis murine model, real-time quantitative PCR expression analysis of liver RNA samples was performed. Compared to the pretreatment level, expression of interleukin-6 was upregulated and that of interleukin 15 receptor alpha was downregulated after the mesenchymal stem cell treatment (unpublished observation).

5. The carbon tetrachloride ($CCl_4$)-induced chronic liver disease model is another model of chronic liver disease (5, 26) that can be used for experimental regenerative therapy. To establish this murine model, C57Bl/6 mice are intraperitoneally injected with 1 ml/kg of $CCl_4$ for 4 weeks. These mice develop advanced fibrotic changes in the liver, i.e., cirrhosis. The therapeutic effect of mesenchymal stem cells on chronic liver disease can also be studied using this model. It is reported that entire fractions of bone marrow cells can improve liver fibrosis in this $CCl_4$-induced cirrhotic murine model, presumably via the activation of matrix metalloproteinase (5). The rat is an alternative rodent for establishing chronic liver injury models, either $CCl_4$-induced (27, 28) or steatohepatitis (29) cirrhosis.

## References

1. Williams R (2006) Global challenges in liver disease. Hepatology;44:521–526.

2. Michalopoulos GK, DeFrances MC (1997) Liver regeneration. Science;276:60–66.

3. Evarts RP, Nagy P, Nakatsukasa H et al. (1989) In vivo differentiation of rat liver oval cells into hepatocytes. Cancer Res;49:1541–1547.

4. Dorrell C, Grompe M (2005) Liver repair by intra- and extrahepatic progenitors. Stem Cell Rev;1:61–64.

5. Sakaida I, Terai S, Yamamoto N et al. (2004) Transplantation of bone marrow cells reduces CCl4-induced liver fibrosis in mice. Hepatology;40:1304–1311.

6. Terai S, Ishikawa T, Omori K et al. (2006) Improved liver function in patients with liver cirrhosis after autologous bone marrow cell infusion therapy. Stem Cells;24:2292–2298.

7. Fujii H, Hirose T, Oe S et al. (2002) Contribution of bone marrow cells to liver regeneration after partial hepatectomy in mice. J Hepatol;36:653–659.

8. Petersen BE, Bowen WC, Patrene KD et al. (1999) Bone marrow as a potential source of hepatic oval cells. Science;284:1168–1170.

9. Jang YY, Collector MI, Baylin SB et al. (2004) Hematopoietic stem cells convert into liver cells within days without fusion. Nat Cell Biol;6:532–539.

10. Lagasse E, Connors H, Al-Dhalimy M et al. (2000) Purified hematopoietic stem cells can differentiate into hepatocytes in vivo. Nat Med;6:1229–1234.

11. Shi XL, Qiu YD, Wu XY et al. (2005) In vitro differentiation of mouse bone marrow mononuclear cells into hepatocyte-like cells. Hepatol Res;31:223–231.

12. Wagers AJ, Sherwood RI, Christensen JL et al. (2002) Little evidence for developmental plasticity of adult hematopoietic stem cells. Science;297:2256–2259.

13. Rosen ED, MacDougald OA (2006) Adipocyte differentiation from the inside out. Nat Rev Mol Cell Biol;7:885–896.

14. Planat-Benard V, Menard C, Andre M et al. (2004) Spontaneous cardiomyocyte differentiation from adipose tissue stroma cells. Circ Res;94:223–229.

15. Gaustad KG, Boquest AC, Anderson BE et al. (2004) Differentiation of human adipose tissue stem cells using extracts of rat cardiomyocytes. Biochem Biophys Res Commun;314:420–427.

16. Banas A, Teratani T, Yamamoto Y et al. (2009) Rapid hepatic fate specification of adipose-derived stem cells and their therapeutic potential for liver failure. J Gastroenterol Hepatol;24:70–77.

17. Banas A, Teratani T, Yamamoto Y et al. (2007) Adipose tissue-derived mesenchymal stem cells as a source of human hepatocytes. Hepatology;46:219–228.

18. Sato Y, Araki H, Kato J et al. (2005) Human mesenchymal stem cells xenografted directly to rat liver are differentiated into human hepatocytes without fusion. Blood;106:756–763.

19. Seo MJ, Suh SY, Bae YC et al. (2005) Differentiation of human adipose stromal cells into hepatic lineage in vitro and in vivo. Biochem Biophys Res Commun;328:258–264.

20. Fang B, Shi M, Liao L et al. (2004) Systemic infusion of FLK1(+) mesenchymal stem cells ameliorate carbon tetrachloride-induced liver fibrosis in mice. Transplantation;78:83–88.

21. Parekkadan B, van Poll D, Megeed Z et al. (2007) Immunomodulation of activated hepatic stellate cells by mesenchymal stem cells. Biochem Biophys Res Commun;363:247–252.

22. Gonzalez MA, Gonzalez-Rey E, Rico L et al. (2009) Adipose-derived mesenchymal stem cells alleviate experimental colitis by inhibiting inflammatory and autoimmune responses. Gastroenterology;136:978–989.

23. Matsuzawa N, Takamura T, Kurita S et al. (2007) Lipid-induced oxidative stress causes steatohepatitis in mice fed an atherogenic diet. Hepatology;46:1392–1403.

24. Tropel P, Noel D, Platet N et al. (2004) Isolation and characterisation of mesenchymal stem cells from adult mouse bone marrow. Exp Cell Res;295:395–406.

25. Soleimani M, Nadri S (2009) A protocol for isolation and culture of mesenchymal stem cells from mouse bone marrow. Nat Protoc;4:102–106.

26. Natsume M, Tsuji H, Harada A et al. (1999) Attenuated liver fibrosis and depressed serum albumin levels in carbon tetrachloride-treated IL-6-deficient mice. J Leukoc Biol;66:601–608.

27. Oyagi S, Hirose M, Kojima M et al. (2006) Therapeutic effect of transplanting HGF-treated bone marrow mesenchymal cells into CCl4-injured rats. J Hepatol;44:742–748.

28. Hardjo M, Miyazaki M, Sakaguchi M et al. (2009) Suppression of carbon tetrachloride-induced liver fibrosis by transplantation of a clonal mesenchymal stem cell line derived from rat bone marrow. Cell Transplant;18:89–99.

29. Ota T, Takamura T, Kurita S et al. (2007) Insulin resistance accelerates a dietary rat model of nonalcoholic steatohepatitis. Gastroenterology;132:282–293.

# INDEX

Takahiro Ochiya (ed.), *Liver Stem Cells: Methods and Protocols*, Methods in Molecular Biology, vol. 826,
DOI 10.1007/978-1-61779-468-1, © Springer Science+Business Media, LLC 2012